International Series in Advanced Management Studies

Editor-in-Chief

Alberto Pastore, Sapienza University of Rome
Rome, Italy

Series Editors

Giovanni Battista Dagnino, University of Rome LUMSA
Palermo, Italy

Marco Frey, Sant'Anna School of Advanced Studies
Pisa, Italy

Christian Grönroos, Hanken School of Economics
Helsinki, Finland

Michael Haenlein, ESCP Europe
Paris, France

Charles F. Hofacker, Florida State University
Tallahassee, USA

Anne Huff, Maynooth University
Maynooth, Ireland

Morten Huse, BI Norwegian Business School
Oslo, Norway

Gennaro Iasevoli, Lumsa University
Rome, Italy

Andrea Moretti, University of Udine
Udine, Italy

Fabio Musso, University of Urbino
Urbino, Italy

Mustafa Ozbilgin, Brunel University London
Uxbridge, UK

Paolo Stampacchia, University of Naples Federico II
Naples, Italy

Luca Zanderighi, University of Milan
Milan, Italy

Assistant Editor

Michela Matarazzo, Marconi University
Rome, Italy

This series is published in cooperation with the Italian Society of Management (SIMA). It is home to books that focus on advanced and cutting-edge management topics that are likely to have significant impacts on and advantages for companies, institutions and other organizations in the business world. The topics reflect various aspects of management – such as business strategy, corporate finance, entrepreneurship, operations, SME, corporate governance, innovation management, marketing, and corporate communication – as applied to all industries, all types of organizations (profit, public or non profit), and all regions of the globe.

This is a SCOPUS indexed book series.

Book selection procedure:

This series follows a single-blind reviewing procedure. Authors interested in publishing a volume are invited to submit their proposal by e-mail to the Editor-in-Chief, Prof. Alberto Pastore: alberto.pastore@uniroma1.it. Authors are requested to supply the following material: structured abstract (and methodology, if appropriate), table of contents, 2 or 3 sample chapters, CV (including a list of previous publications).

Francesca Romana Arduino

Governance and Sustainability of Family Business

Evidence from Domestically Listed Italian Companies

Francesca Romana Arduino
Department of Human Sciences
European University of Rome
Rome, Italy

ISSN 2366-8814 ISSN 2366-8822 (electronic)
International Series in Advanced Management Studies
ISBN 978-3-032-01943-1 ISBN 978-3-032-01944-8 (eBook)
https://doi.org/10.1007/978-3-032-01944-8

© The Editor(s) (if applicable) and The Author(s), under exclusive license to Springer Nature Switzerland AG 2025

This work is subject to copyright. All rights are solely and exclusively licensed by the Publisher, whether the whole or part of the material is concerned, specifically the rights of translation, reprinting, reuse of illustrations, recitation, broadcasting, reproduction on microfilms or in any other physical way, and transmission or information storage and retrieval, electronic adaptation, computer software, or by similar or dissimilar methodology now known or hereafter developed.
The use of general descriptive names, registered names, trademarks, service marks, etc. in this publication does not imply, even in the absence of a specific statement, that such names are exempt from the relevant protective laws and regulations and therefore free for general use.
The publisher, the authors and the editors are safe to assume that the advice and information in this book are believed to be true and accurate at the date of publication. Neither the publisher nor the authors or the editors give a warranty, expressed or implied, with respect to the material contained herein or for any errors or omissions that may have been made. The publisher remains neutral with regard to jurisdictional claims in published maps and institutional affiliations.

This Springer imprint is published by the registered company Springer Nature Switzerland AG
The registered company address is: Gewerbestrasse 11, 6330 Cham, Switzerland

If disposing of this product, please recycle the paper.

Competing Interests The author has no competing interests to declare that are relevant to the content of this manuscript.

Contents

1	**Introduction**...	1
	References..	3
2	**Corporate Social Responsibility and Sustainability**...............	5
	2.1 Overview and Definition.................................	5
	2.2 Stakeholder Theory.....................................	8
	2.3 Who Promotes Sustainability in Business?.................	10
	2.4 The Rise of ESG..	13
	2.5 Combining Profit and Purpose...........................	15
	References..	17
3	**Governance and Sustainability Practices in Family Businesses**......	21
	3.1 Relationship Between Governance and Sustainability.........	21
	3.2 Understanding Sustainability in the Family Business Context.....	23
	3.3 How to Foster Sustainable Governance......................	25
	3.4 Governance and Sustainability in Italy.....................	27
	References..	31
4	**Management and Governance of Family Business**.................	37
	4.1 Overview of the Family Business Phenomenon................	37
	4.2 Definition and Distinctive Features.........................	41
	4.3 Main Theoretical Frameworks.............................	45
	4.3.1 Agency Theory..................................	46
	4.3.2 Resource-based View Theory.......................	48
	4.3.3 Socioemotional Wealth Theory.....................	50
	4.3.4 Stewardship Theory..............................	52
	4.4 Corporate Governance in Family Business....................	53
	References..	57
5	**Empirical Analysis of Italian Listed Family Businesses**............	65
	5.1 Objectives and Scope of the Study.........................	65
	5.2 Methodology and Data Sources...........................	65

	5.3 Sample	73
	5.4 Comparative Analysis of Governance and Sustainability Practices	75
	5.4.1 Corporate Governance Characteristics	75
	5.4.2 Sustainability Characteristics	77
	5.5 Best Practices and Lessons Learned	87
	References	89
6	**Conclusion**	91
	6.1 Summary of Key Findings	91
	6.2 Implications for Family Businesses	92
	6.3 Policy Recommendations	93
	6.4 Final Thoughts	94
	References	94

Chapter 1
Introduction

Over the last few years, attention to sustainability has increased notably; this change is driven primarily by policymakers, regulators, investors, consumers, and other stakeholders and encourages companies to prioritize sustainability-oriented actions in their strategic decision-making (Weaver et al., 1999).

From a regulatory perspective, this increased focus has led to directives and guidelines being issued across numerous countries with the goal of fostering sustainability awareness and encouraging firms to implement sustainable initiatives. The European Union (EU) in particular has taken a proactive stance in promoting sustainability-related practices by enacting several laws, such as the Nonfinancial Reporting Directive (NFRD), known as Directive 2014/95/EU, which requires large public interest companies to disclose nonfinancial information on environmental, social, and governance (ESG) matters and the Corporate Sustainability Reporting Directive (CSRD), which expands the scope to include more companies and mandates more detailed and standardized sustainability reporting. The EU has also recently enacted the Corporate Sustainability Due Diligence Directive (CSDDD), the goal of which is to ensure that companies operating within the EU identify, prevent, mitigate, and account for adverse human rights and environmental impacts in their operations, subsidiaries, and value chains.

Although the achievement of good financial performance is still imperative for corporate growth, increasing evidence suggests that the development of a corporate social responsibility (CSR) strategy that effectively targets a variety of goals and addresses the demands of diverse stakeholders can provide companies with significant long-term benefits, including improved stakeholder relationships; ultimately, this can contribute to sustainable business success.

Since the emphasis on sustainability is growing, companies have begun to update their governance structures to be better aligned with sustainability objectives from a corporate governance perspective.

© The Author(s), under exclusive license to Springer Nature Switzerland AG 2025
F. R. Arduino, *Governance and Sustainability of Family Business*, International Series in Advanced Management Studies,
https://doi.org/10.1007/978-3-032-01944-8_1

The transition toward sustainable corporate governance is particularly important in the landscape of family businesses. Family firms represent the predominant organizational form worldwide (La Porta et al., 1999) and thus exert a relevant influence on the global economy (Astrachan & Shanker, 2003; Morck & Yeung, 2004). Hence, understanding whether family firms are more responsive than nonfamily firms to environmental and social concerns has far-reaching implications.

Because of their unique characteristics, family businesses can be better positioned than nonfamily businesses to implement their sustainability strategies (Ferreira et al., 2021). According to the socioemotional wealth (SEW) perspective (Gómez-Mejía et al., 2007), family businesses exhibit a strong inclination toward nonfinancial objectives that satisfy the family's emotional needs, such as preserving the family legacy (Berrone et al., 2012).

The binding social ties family firms experience further strengthen this inclination, as the close relationships within the family and with key stakeholders promote a commitment to values that prioritize long-term goals (Berrone et al., 2012).

Sustainability initiatives inherently resonate with these objectives by ensuring long-term viability and reputation, which are crucial for maintaining family cohesion and reputation across generations—key dimensions of socioemotional wealth. In particular, the extended SEW perspective (Miller & Le Breton-Miller, 2014) emphasizes how family firms often prioritize values such as legacy, community engagement, and moral stewardship, viewing sustainability as not only a strategic opportunity, but also a reflection of their identity and desire to preserve the family's positive image and influence over time.

In this work, the analysis focuses on a sample of Italian family firms listed on the Milan Stock Exchange. Italy is a country that is characterized by a vast diffusion of companies controlled by a family or a coalition of families, representing 65% of the total number of Italian companies in 2022, and the Milan Stock Exchange represents the highest Euronext market in terms of incidence of family-controlled companies (74.9%) (AIDAF, 2024). Thus, this study aims to investigate the main characteristics of Italian listed family firms in terms of governance and sustainability performance, building upon the extant literature on this topic.

This book is structured as follows. Chapter 2 lays the foundation for understanding CSR and sustainability. It begins with an overview and definition of these concepts, explores stakeholder theory, and identifies key promoters of sustainability in business. Moreover, it examines the emergence of ESG criteria and the rise of B Corps and benefit corporations as models for sustainable business practices. Chapter 3 explores how sustainability is implemented in family businesses. Specifically, it investigates the relationship between governance structures and sustainability outcomes, providing insights into the ways in which effective governance can foster sustainable business practices. Chapter 4 provides an in-depth examination of family businesses, starting with an overview and definition and highlighting their distinctive features. It reviews major theoretical frameworks in the literature, such as agency theory, resource-based view theory, SEW theory, and stewardship theory, and it concludes by examining the influence of family dynamics on the corporate governance practices of family businesses. Chapter 5 presents the objectives, methodology, and data sources of a study of a sample of Italian domestically listed family firms. This chapter includes a comparative analysis of the governance and sustainability practices of these firms, highlighting best practices and lessons gleaned from the empirical data. The final Chap. 6 summarizes the key findings of the book, discusses their implications for family businesses, and provides policy recommendations. Finally, the book concludes with final considerations on the evolving role of governance in promoting sustainability in family businesses.

References

AIDAF. (2024). *Family businesses in Italy*. Retrieved July 1, 2024, from https://www.aidaf.it/en/family-businesses/

Astrachan, J. H., & Shanker, M. C. (2003). Family businesses' contribution to the U.S. economy: A closer look. *Family Business Review, 16*(3), 211–219. https://doi.org/10.1177/08944865030160030601

Berrone, P., Cruz, C., & Gomez-Mejia, L. R. (2012). Socioemotional wealth in family firms: Theoretical dimensions, assessments approaches and agenda for future research. *Family Business Review, 25*(3), 258–279. https://doi.org/10.1177/0894486511435355

Ferreira, J. J., Fernandes, C. I., Schiavone, F., & Mahto, R. V. (2021). Sustainability in family business–A bibliometric study and a research agenda. *Technological Forecasting and Social Change, 173*, 121077. https://doi.org/10.1016/j.techfore.2021.121077

Gómez-Mejía, L. R., Haynes, K. T., Núñez-Nickel, M., Jacobson, K. J. L., & Moyano-Fuentes, J. (2007). Socioemotional wealth and business risks in family-controlled firms: Evidence from Spanish olive oil mills. *Administrative Science Quarterly, 52*(1), 106–137. https://doi.org/10.2189/asqu.52.1.106

La Porta, R., Lopez-De-Silanes, F., & Shleifer, A. (1999). Corporate ownership around the world. *The Journal of Finance, 54*(2), 471–517. https://doi.org/10.1111/0022-1082.00115

Miller, D., & Le Breton-Miller, I. (2014). Deconstructing socioemotional wealth. *Entrepreneurship Theory and Practice, 38*(4), 713–720. https://doi.org/10.1111/etap.12111

Morck, R., & Yeung, B. (2004). Family control and the rent-seeking society. *Entrepreneurship Theory and Practice, 28*(4), 391–409. https://doi.org/10.1111/j.1540-6520.2004.00053.x

Weaver, G. R., Trevino, L. K., & Cochran, P. L. (1999). Corporate ethics programs as control systems: Influences of executive commitment and environmental factors. *Academy of Management Journal, 42*(1), 41–57. https://doi.org/10.2307/256873

Chapter 2
Corporate Social Responsibility and Sustainability

2.1 Overview and Definition

In recent centuries, the development of corporations in modern society has given rise to ongoing discussions about their primary objectives, particularly with respect to which interests should take precedence in their strategic decision-making. To address this issue, two distinct approaches—the shareholder approach and the stakeholder approach—have emerged, each providing unique viewpoints.

The shareholder value creation approach is centered around the primary objective of maximizing the wealth of shareholders, who are the owners of a company. In this approach, strategic decisions are evaluated primarily on the basis of their impact on shareholder returns; however, this often leads to an emphasis on short-term financial performance at the expense of long-term sustainability and ethical considerations. In contrast, the goal of the stakeholder value creation approach is to create value for all of the stakeholders involved in or affected by the company's operations; these may include shareholders, employees, suppliers, customers, local communities, and the environment. When this approach is utilized, strategic decisions are evaluated on the basis of their impact on all stakeholders, with a focus on balancing various interests. With this approach, long-term sustainability and the implementation of responsible business practices tend to be emphasized.

The ongoing debate about the objectives that companies should prioritize has been reignited in recent decades, raising the question of whether a company can have responsibilities that extend beyond those that are purely economic and financial.

According to McGuire et al. (1988), *"the idea of social responsibilities supposes that the corporation has not only economic and legal obligations but also certain responsibilities to society which extend beyond these obligations."* From this perspective, the role of corporations in society transcends mere profit generation. By adopting social responsibilities, companies can contribute to societal well-being and environmental protection, enhance their sustainability impact, and finally, create a positive cycle of mutual benefit between business and society.

CSR was first introduced in 1953 by Howard Bowen in his seminal book, *Social Responsibilities of the Businessman* (Bowen, 1953). In this book, Bowen emphasized the importance of considering the interests of not only shareholders but also a diverse range of stakeholders, including employees, suppliers, customers, and society as a whole, while adopting a long-term perspective. However, it took several years for this concept to be widely accepted and integrated into corporate practices. Although, in 1970, Milton Friedman notoriously affirmed in the *New York Time Magazine* that the "*social responsibility of business is to increase its profits*" (Friedman, 1970), the idea of corporate social and environmental responsibility continued to spread, and since the end of the twentieth century and the beginning of the twenty-first century, interest in issues of sustainability has increased dramatically.

During the 1960s and 1970s, the civil rights movement, environmental activism, and consumer protection efforts increased awareness of the impacts of corporations on society. During this period, calls for businesses to adopt more responsible practices increased.

In the 1980s and 1990s, CSR began to gain more traction as companies began to recognize the potential benefits of ethical practices, such as enhanced reputation, customer loyalty, and risk management. The introduction of the triple bottom-line concept, in which a company's success is measured on the basis of economic, social, and environmental performance, further consolidated the importance of CSR.

According to Elkington (1999), companies should commit to focusing on social and environmental concerns in the same way as they do for economic concerns and measure their success on three pillars: people, the planet, and profit.

In the 2000s, with globalization and the rise of digital communication, the need for transparency increased, encouraging more companies to adopt CSR initiatives. Major corporate scandals and environmental disasters also highlighted the need for responsible corporate behavior.

The United Nations (UN) also played an important role in this transition, as in 2000, it introduced the UN Global Compact, a voluntary program aimed at encouraging companies to adopt ethical and sustainable practices. By participating in the UN Global Compact, organizations commit to embedding some guiding principles into their corporate culture, strategic planning, and daily operations. These principles are related to critical areas such as human rights, labor standards, environmental protection, and anticorruption efforts. The Millennium Development Goals (MDGs) (United Nations Millennium Development Goals, 2000) are a set of eight international development goals that were established following the Millennium Summit of the United Nations in 2000. All UN member states agreed upon the goals, which aimed to address a wide range of global challenges by 2015. Succeeding the MDGs, the Sustainable Development Goals (SDGs) were adopted by all UN member states in 2015. The SDGs are a set of 17 goals that are designed to be a "blueprint to achieve a better and more sustainable future for all" (United Nations Sustainable Development Goals, 2015) by 2030. These goals address a broad range

2.1 Overview and Definition

Table 2.1 The United Nations sustainable development goals

Goal	Description
No poverty	End poverty in all its forms everywhere.
Zero hunger	End hunger, achieve food security and improved nutrition, and promote sustainable agriculture.
Good health and well-being	Ensure healthy lives and promote well-being for all at all ages.
Quality education	Ensure inclusive and equitable quality education and promote lifelong learning opportunities for all.
Gender equality	Achieve gender equality and empower all women and girls.
Clean water and sanitation	Ensure the availability and sustainable management of water and sanitation for all.
Affordable and clean energy	Ensure access to affordable, reliable, sustainable, and modern energy for all.
Decent work and economic growth	Promote sustained, inclusive, and sustainable economic growth, full and productive employment, and decent work for all.
Industry, innovation, and infrastructure	Build resilient infrastructure, promote inclusive and sustainable industrialization, and foster innovation.
Reduced inequality	Reduce inequality within and among countries.
Sustainable cities and communities	Make cities and human settlements inclusive, safe, resilient, and sustainable.
Responsible consumption and production	Ensure sustainable consumption and production patterns.
Climate action	Take urgent action to combat climate change and its impacts.
Life below water	Conserve and sustainably use the oceans, seas, and marine resources for sustainable development.
Life on land	Protect, restore, and promote the sustainable use of terrestrial ecosystems, manage forests sustainably, combat desertification, halt and reverse land degradation, and halt biodiversity loss.
Peace, justice, and strong institutions	Promote peaceful and inclusive societies for sustainable development, provide access to justice for all, and build effective, accountable, and inclusive institutions at all levels.
Partnerships for the goals	Strengthen the means of implementation and revitalize the global partnership for sustainable development.

Source: United Nations. Adapted from https://sdgs.un.org/goals

of interconnected global challenges, including those related to poverty, inequality, climate change, peace, and justice, as described in Table 2.1; thus, these goals inform an ambitious global agenda.

In 2019, the Business Roundtable statement on corporate purpose marked a significant shift from the traditional shareholder-centric model to a stakeholder-centric model of corporate governance. This statement, which was signed by the CEOs of nearly 200 of the largest companies in the U.S., redefines the purpose of a corporation and declares that companies should deliver long-term value to all stakeholders, including customers, employees, suppliers, and communities, not just shareholders, thus promoting a more inclusive and sustainable economic system.

Sustainability is not only currently receiving increased attention from governments and institutions but has also become an integral part of corporate strategy for many businesses, encompassing a wide range of practices. Additionally, investors increasingly evaluate companies not only in terms of their financial performance but also in terms of their social and environmental contributions, reflecting a broader understanding of their role in society.

2.2 Stakeholder Theory

Stakeholder theory represents a significant shift in the understanding of corporate responsibility and governance. Specifically, this theory promotes a more inclusive and sustainable approach to business by extending the focus from creating value for only shareholders to including all stakeholders.

The early roots of stakeholder theory can be traced back to the 1960s and 1970s, when scholars began to question the dominance of the shareholder value creation approach. During this period, new perspectives about CSR began to gain momentum, calling upon companies to also consider societal and environmental dimensions.

The term "stakeholder" first appeared in 1963 in the Stanford Memo, a report by the Stanford Research Institute, in which stakeholders were identified as "those groups without whose support the organization would cease to exist" (see Mitchell et al., 1997, p. 858), thus suggesting that the dependence of companies on stakeholders be used as a rationale for identifying the latter.

Stakeholder theory was later formalized by R. Edward Freeman in 1984 in his seminal book *Strategic Management: A Stakeholder Approach* (Freeman, 1984). According to Freeman, companies should create value not only for their shareholders but also for all of their stakeholders. He identified stakeholders as "any group or individual who can affect or is affected by the achievement of the organization's objectives" (Freeman, 1984, p. 46), that is, employees, suppliers, customers, communities, and financial institutions. Since the 1980s, the importance of stakeholder theory has grown significantly, with scholars increasingly investigating the sustainability of prioritizing shareholder wealth as the primary goal of companies (Berman et al., 1999; Donaldson & Preston, 1995; Freeman et al., 2004; Hillman & Keim, 2001; Jensen, 2001; Mitchell et al., 1997). Moreover, stakeholder theory has been expanded across many studies to include ethics, management, and governance considerations. Specifically, this theory has been integrated with concepts from business ethics, as companies are considered to have moral obligations to consider the rights and interests of all of their stakeholders (Buchholtz & Carroll, 2014; Jones, 1995).

On the basis of this theory, companies should identify and map all relevant stakeholders who can affect or are affected by the organization (see Fig. 2.1). Notably, companies are expected to understand and balance the diverse interests and expectations of various stakeholders and need to be able to recognize their power and influence in shaping organizational strategies and outcomes.

2.2 Stakeholder Theory

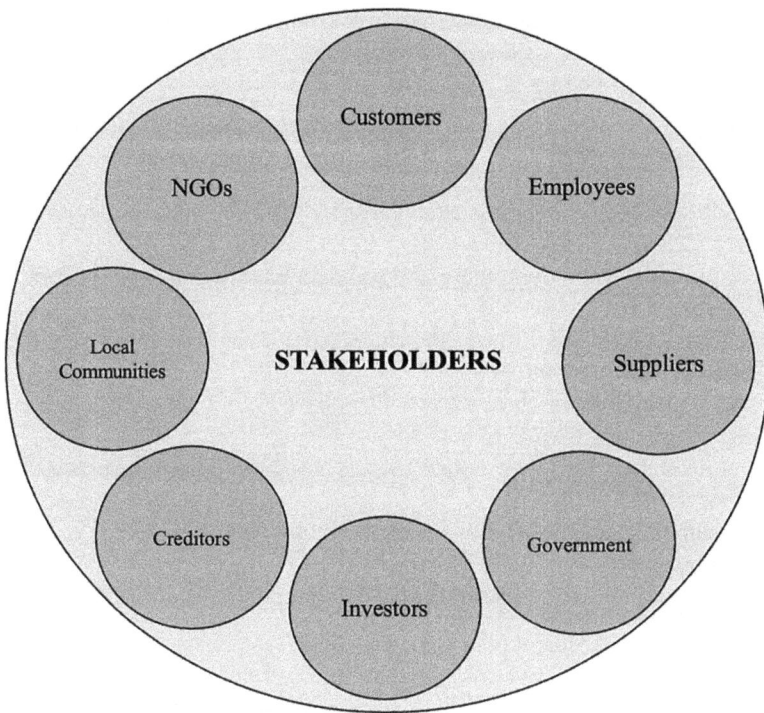

Fig. 2.1 Types of company stakeholders. Source: Author's elaboration

With respect to the stakeholder definition, many possible classifications have been proposed, including the following:

- Primary versus secondary stakeholders (Clarkson, 1995): primary stakeholders have a direct, formal, or contractual relationship with an organization, and they are essential for the organization's survival and success (e.g., employees, customers, suppliers, shareholders, and creditors), whereas secondary stakeholders do not have a direct contractual relationship with the organization but can influence or be influenced by the company's actions (e.g., media, advocacy groups, trade associations, and the general public).
- Voluntary versus involuntary stakeholders (Clarkson, 1998): voluntary stakeholders choose to be involved with an organization and have a mutual relationship with it (e.g., employees, investors, customers, and suppliers), whereas involuntary stakeholders are affected by the organization's actions even though they did not choose to be involved (e.g., local communities and the general public).
- Internal versus external stakeholders (Carroll & Näsi, 1997): internal stakeholders are individuals or groups within an organization who are directly involved in its operations and governance (e.g., shareholders, managers, and employees), whereas external stakeholders are individuals or groups outside the organization

who are affected by or can affect the organization's activities (e.g., customers, suppliers, competitors, creditors, governments, communities, NGOs, media, and the natural environment).

Additionally, according to Mitchell et al. (1997), stakeholders can be classified on the basis of the attributes of power, urgency, and legitimacy as follows:

- Dormant stakeholders: these stakeholders have power but lack legitimacy and urgency.
- Discretionary stakeholders: these stakeholders have legitimacy but lack power and urgency.
- Demanding stakeholders: these stakeholders have urgency but lack power and legitimacy.
- Dominant stakeholders: these stakeholders have both power and legitimacy but not urgency.
- Dangerous stakeholders: these stakeholders have both power and urgency but not legitimacy.
- Dependent stakeholders: these stakeholders have both legitimacy and urgency but not power.
- Definitive stakeholders: these stakeholders have power, legitimacy, and urgency.

However, the management of diverse stakeholder interests can be complex, challenging and resource intensive for companies; in fact, potential conflicts may arise between different types of stakeholders with competing interests. Therefore, companies are increasingly involved in stakeholder engagement activities to stimulate fruitful dialog and promote trust and collaboration with diverse stakeholders (Morsing & Schultz, 2006; Sen et al., 2006). Studies on stakeholder engagement practices have demonstrated their positive impact on corporate decision-making and organizational outcomes, including learning and knowledge creation (Desai, 2018; Mitchell et al., 2022), highlighting that stakeholder engagement is relevant to value creation (Freudenreich et al., 2020; Harrison & Wicks, 2013).

2.3 Who Promotes Sustainability in Business?

Over the last few decades, the growing interest in sustainability has been driven by external and internal organizational drivers, and various stakeholders have increasingly contributed to raising awareness of these issues.

International organizations, such as the UN, the Organization for Economic Co-operation and Development (OECD), and the International Labour Organization (ILO), have been instrumental in increasing awareness of sustainability and encouraging responsible and ethical practices. By engaging countries in a collective effort to address global challenges, these organizations have made significant progress in this area. Specifically, the introduction of the 17 ambitious SDGs by the UN, which need to be reached by 2030, has served as a driving force for transformative action

2.3 Who Promotes Sustainability in Business?

among all UN member states. Countries are encouraged to adopt and implement sustainable practices that promote economic growth, social inclusion, and environmental protection through the alignment of their national policies with these global objectives. Thus, the UN's efforts to promote international cooperation and provide a common framework have been pivotal in driving global sustainability initiatives.

In 2006, the United Nations Environment Programme Finance Initiative (UNEP FI) introduced the Principles for Responsible Investment (PRI) (United Nations Principles for Responsible Investment, 2006), illustrated in Table 2.2. The goal of this innovative initiative was to incorporate ESG factors into investment decision-making processes, thereby promoting sustainable and responsible investment practices worldwide. The PRI provided a voluntary framework that investors can utilize to integrate ESG considerations into their portfolios and investment practices. By adhering to these principles, investors commit to acting in the best long-term interests of their beneficiaries and the broader society: "As institutional investors, we have a duty to act in the best long-term interests of our beneficiaries. In this fiduciary role, we believe that environmental, social, and corporate governance (ESG) issues can affect the performance of investment portfolios (to varying degrees across companies, sectors, regions, asset classes and through time)" (United Nations Principles for Responsible Investment, 2006). Thus, the launch of the PRI marked a significant milestone in the global movement toward sustainable finance, encouraging investors to consider the long-term impacts of their investments on society and the environment. Since its institution, the PRI have gained widespread acceptance and have been endorsed by numerous institutional investors, asset managers, and financial institutions, thus significantly influencing the global investment landscape toward greater sustainability and ethical responsibility.

Institutional investors have become increasingly active and engaged in incorporating ESG considerations into their investment decisions. Recognizing the long-term value of sustainable practices, these investors now prioritize companies that have strong ESG performance. This shift is evident in the rise of sustainable

Table 2.2 Principles for responsible investment

Principle	Description
1	We will incorporate ESG issues into investment analysis and decision-making processes.
2	We will be active owners and incorporate ESG issues into our ownership policies and practices.
3	We will seek appropriate disclosure on ESG issues by the entities in which we invest.
4	We will promote acceptance and implementation of the principles within the investment industry.
5	We will work together to enhance our effectiveness in implementing the principles.
6	We will each report on our activities and progress toward implementing the principles.

Notes: ESG denotes Environmental, Social and Governance
Source: United Nations Principles for Responsible Investment. Adapted from https://www.unpri.org/

investment funds and ESG-focused indices, such as the FTSE4Good Index, the Dow Jones Sustainability Index (DJSI), and the MSCI Global Sustainability Index. For companies to be included and continue to be present in these stock indices, they need to make progress in sustainability matters, which also affects their eligibility for consideration in investment portfolios. Institutional investors may use their significant financial influence to encourage corporate transparency and accountability, and they may even divest from companies that fail to meet ESG criteria or use their voting power to advocate for better practices (Eccles & Klimenko, 2019).

Policymakers and regulators have increasingly prioritized sustainability by developing guidelines and implementing directives, although these have significant differences and regulatory fragmentation across countries. The goals of these initiatives are to increase sustainability awareness, encourage sustainable corporate governance, and pressure companies to increase their ESG reporting. The EU in particular has enacted several laws, including the Non-Financial Reporting Directive (NFRD) (Directive 2014/95/EU) (EU Non-Financial Reporting Directive, 2014), the CSRD (EU Corporate Sustainability Reporting Directive, 2022), and the recent CSDDD (EU Corporate Sustainability Due Diligence Directive, 2024), and has mandated the creation of the EU taxonomy for sustainable activities (EU Taxonomy for Sustainable Activities, 2020), which provides a clear classification system to guide investments toward environmentally sustainable projects. Investors and companies can use this regulatory framework to identify and prioritize activities that contribute to the EU's environmental objectives, such as the European Green Deal (European Commission, 2020). By setting standards and enforcing compliance, regulators ensure that companies integrate sustainability into their operations, thereby driving systemic change across industries.

Civil society, particularly Millennials and Generation Z, demonstrates a stronger commitment to social and environmental responsibility than previous generations have shown. Individuals in these younger groups are more informed about global challenges and prioritize sustainability in their daily lives. Their purchasing behavior reflects this shift; in fact, they prefer to buy from companies that offer sustainable and eco-friendly products (Bedard & Tolmie, 2018; Bucic et al., 2012). This consumer demand encourages businesses to innovate and adopt greener practices to remain competitive. Additionally, social media and digital platforms amplify the voices of these generations, allowing them to advocate for change and demand accountability from companies. Thus, the growing influence of socially and environmentally conscious consumers is reshaping markets and pushing businesses to align with sustainability values (Cohen & Muñoz, 2017).

Millennials are also leading the charge in socially responsible investments. In fact, compared with older generations, younger generations are more likely to consider ESG factors when making investment decisions (Cheah et al., 2011). This demographic shift influences the demand for investment products that align with Millennials' values, such as green bonds and socially responsible mutual funds. Moreover, the increased availability of information on ESG performance enables individual investors to make more informed choices. Financial institutions have responded by offering a broader range of ESG investment options (such as green bonds), meeting the preferences of this socially conscious investor base. As this trend continues, it encourages more companies to adopt sustainable practices to attract and retain investment.

Finally, employees constitute another important stakeholder group that is advancing sustainability within companies (Wolf, 2013). Employees can promote sustainable practices by raising awareness about social and environmental issues, collaborate with management to integrate sustainability into the company's strategy and operations, and contribute to sustainability by developing and implementing innovative solutions beyond the workplace for the broader community.

2.4 The Rise of ESG

ESG represents a set of criteria that are used to evaluate a company's performance in areas that are considered nonfinancial but are critical for long-term sustainability and ethical impact. The term "ESG" was first coined in the early 2000s, particularly in the 2004 report "Who Cares Wins—Connecting Financial Markets to a Changing World" (United Nations Environment Programme Finance Initiative Report, 2004) developed by the UN in collaboration with financial institutions and the Swiss government. This report explicitly called for the better integration of ESG factors into investment analysis and decision-making processes. Table 2.3 illustrates the meaning of the three pillars of the ESG acronym.

The roots of ESG can be traced back to the broader concept of Socially Responsible Investing (SRI), which began to be more widespread in the 1960s and 1970s, when investors started to consider ethical issues in their investment decisions (Martini, 2021). In the 1980s, the CSR movement laid the groundwork for the later development of ESG by emphasizing the importance of ethical conduct and social responsibility (Agudelo et al., 2019). In the 1990s and early 2000s, the formalization of sustainability and responsible investment principles progressed significantly. In fact, in 1997, the Global Reporting Initiative (GRI) (Global Reporting Initiative, 1997) was founded to provide a standardized framework for sustainability

Table 2.3 The three pillars of ESG

Environmental	Social	Governance
Environmental factors include efforts to fight climate change, manage carbon emissions, implement effective waste management practices, and protect biodiversity.	Social factors cover adhering to labor law standards (such as prohibiting child labor and discrimination), ensuring adequate working conditions, respecting human rights, complying with workplace health and safety regulations, promoting community relations, and supporting social commitments.	Governance factors include board composition (e.g., diversity), executive compensation practices, the protection of shareholder rights, measures against bribery and corruption, and the encouragement of whistleblowing.

Source: Author's elaboration

reporting, helping companies measure and communicate their environmental, social, and economic impacts, and in 2000, the UN launched the Global Compact (United Nations Global Compact, 2000), encouraging businesses worldwide to adopt sustainable and socially responsible policies. In 2006, the launch of the PRI (United Nations Principles for Responsible Investment, 2006) supported by the UN was a pivotal moment for ESG, as it provided a framework that investors could use to incorporate ESG factors into their investment practices. The principles were quickly accepted, attracting signatories from major institutional investors worldwide. Since then, ESGs have moved from being niche to being mainstream in the investment community. Financial institutions began developing ESG-specific funds, and ESG indices were created to track the performance of companies meeting certain ESG standards.

Larry Fink, CEO of the global investment management company BlackRock, has been one of the most influential actors promoting stakeholder capitalism and sustainable investing. Beginning with his 2018 letter titled "A Sense of Purpose," Fink urged business leaders to embed sustainability issues into their corporate strategies and encouraged investors to consider environmental, social, and governance factors in their investment decision-making. Larry Fink's letters (Blackrock Larry Fink's Letter to CEOs, 2018) significantly impacted the corporate and investment sectors, highlighting the importance of integrating sustainable considerations into business practices and increasing awareness of these issues.

ESG data providers play a crucial role in ensuring the availability of information about companies' consideration of ESG issues. These providers collect, analyze, and distribute ESG data, offering insights into how well companies are addressing these critical factors. By providing comprehensive ESG ratings and reports, they help investors make informed decisions on the basis of third-party assessments of a company's ESG performance.

As sustainability has become an increasingly central concern in the investment community, the importance of the role of ESG data providers has increased. Specifically, investors depend on the reliability and accuracy of the ratings that these providers supply to evaluate the potential risks and opportunities associated with their investments. This highlights the role of ESG data providers in influencing investment strategies and promoting transparency in corporate practices. There are several types of ESG data providers, which differ depending on their focus, methodologies, and types of data offered; these include the following:

- ESG rating agencies: These agencies evaluate companies on the basis of their ESG performance and provide ratings that reflect how well a company is managing its ESG risks and opportunities. These ratings are often used by investors to compare companies inside an industry (e.g., MSCI ESG Ratings, S&P Global ESG Scores, and Sustainalytics).
- Data aggregators: These aggregators collect ESG data from various sources, including company disclosures, regulatory filings, and third-party reports. They add this information to comprehensive databases that investors and analysts can

access for research and analysis purposes (e.g., Bloomberg ESG Data and LSEG Refinitiv ESG Data).
- Alternative data providers: These providers use unconventional data sources and advanced analytics, such as satellite imagery, social media analysis, and AI, to assess companies' ESG performance. These data providers can offer unique insights and real-time data that traditional data sources might not provide (e.g., RepRisk and Arabesque S-Ray).

Although they have an increasingly important role, ESG ratings have faced growing criticism, mainly due to a lack of standardized methodologies across different ESG data providers. This inconsistency can lead to significant variations in ESG ratings for the same company, reducing data reliability (Dorfleitner et al., 2015). Moreover, some ESG data providers do not fully disclose their methodologies or the criteria that they use to evaluate companies and may be exposed to subjectivity in scoring by weighing factors differently, which may lead to inconsistent ratings (Abhayawansa & Tyagi, 2021). Additionally, some ESG data providers also offer consulting services to the companies they rate, which can create potential conflicts of interest. This dual role may raise questions about the impartiality and objectivity of their ratings (Financial Times, 2023).

Finally, the quality and accuracy of ESG data can be questioned, particularly when providers rely on self-reported information from companies. Companies may engage in "greenwashing" practices by overstating their sustainability efforts to appear more sustainable than they truly are, which may lead to misleading and deceptive ratings (Yu et al., 2020).

2.5 Combining Profit and Purpose

When fully embraced, sustainability can be a transformative force for good. The emergence of purpose-driven business models, which strive to achieve positive economic, social, and environmental outcomes, responds to the growing demand for businesses to promote positive societal change. However, companies face complex challenges in embedding sustainability principles into their operations (Birkinshaw et al., 2014).

Many companies struggle with ethical dilemmas and the challenge of balancing profit-driven goals with social and environmental responsibilities. Some pioneering companies have redefined their business missions by formally committing to sustainability-related objectives, such as benefit corporations and Certified B Corporations.

In recent years, various countries have introduced "profit-with-purpose" corporate forms as an innovative legal business form in their corporate law (Levillain & Segrestin, 2019). The first country that introduced new legal forms of corporation was the United States, where three new types of business forms—benefit corporations, social purpose corporations, and public benefit corporations—have been established since 2010. Italy followed suit in 2016, introducing the Società Benefit statute, and in 2019, France instituted a similar model, the Société à Mission. The introduction of these innovative purpose-driven legal business forms has the potential to significantly transform and renew corporate governance practices. These corporations, in addition to generating value for shareholders, are legally bound to integrate social or environmental objectives into their articles of incorporation, as their corporate purpose is to generate a material positive impact on society and the environment. This determines a legal commitment for both shareholders and the board of directors, whose duties are extended to consider other interests in addition to shareholders' financial interests (Clark & Babson, 2012; Hiller, 2013).

The motivations for adopting a "profit-with-purpose" model can differ. According to George et al. (2023), the primary factors driving this shift toward a purpose-driven approach include internal drivers, such as the vision, values, and beliefs of the organizational leaders, typically the founders, or high levels of employee engagement, and external drivers, such as market expectations and the influence of social movements.

Another approach that highlights a firm's commitment to a governance philosophy that differs from the traditional shareholder-centric model is the decision to undergo a rigorous certification process to become a Certified B Corporation (B Corp) (Kim et al., 2016). To receive a B Corp certification (B Corporation, 2024) by the nonprofit organization B Lab, companies have to pass a certain performance threshold (i.e., 80/200) on different dimensions related to how they create value for nonshareholding stakeholders (i.e., employees, customers, suppliers, local communities, and the environment). Specifically, to receive the certification, companies must do the following (B Corporation Certification, 2024):

- "Demonstrate high social and environmental performance by achieving a B Impact Assessment score of 80 or above and passing the risk review. Multinational corporations must also meet baseline requirement standards.
- Make a legal commitment by changing their corporate governance structure to be accountable to all stakeholders, not just shareholders, and achieve benefit corporation status if available in their jurisdiction.
- Exhibit transparency by allowing information about their performance measured against B Lab's standards to be publicly available on their B Corp profile on B Lab's website."

As businesses face growing pressure and an increasingly complex regulatory environment, new B Corp standards have been designed and are set to be introduced in 2026. Currently, more than 10,000 Certified B Corps, ranging from small to large, operate across more than 100 countries and spanning more than 160 industries worldwide.

A significant factor behind the rise of B Corps is the growing effort by profit-driven companies to portray themselves as "green" and "good" (Kim et al., 2016). In this way, these companies aim to better communicate their commitment to society and the natural environment.

According to Gehman et al. (2019), the main value of certification lies not in its potential for organizational growth but in the process itself, which helps organizations better understand stakeholder needs and improves their response to them.

Innovative models such as Benefit Corporations and B Corps offer businesses the opportunity to embrace their role as agents of positive societal change. By striving to combine profit and purpose objectives, these models contribute to the broader discourse on reinventing capitalism to be more humane and sustainable (Marquis, 2020).

References

Abhayawansa, S., & Tyagi, S. (2021). Sustainable investing: The black box of environmental, social, and governance (ESG) ratings. *The Journal of Wealth Management, 24*(1), 49–54. https://doi.org/10.3905/jwm.2021.1.130

Agudelo, M. A. L., Jóhannsdóttir, L., & Davídsdóttir, B. (2019). A literature review of the history and evolution of corporate social responsibility. *International Journal of Corporate Social Responsibility, 4*(1), 1–23. https://doi.org/10.1186/s40991-018-0039-y

B Corporation. (2024). *Make business a force for good*. Retrieved July 6, 2024, from https://www.bcorporation.net/en-us/

B Corporation Certification. (2024). *Measuring a company's entire social and environmental impact*. Retrieved July 6, 2024, from https://www.bcorporation.net/en-us/certification/

Bedard, S. A. N., & Tolmie, C. R. (2018). Millennials' green consumption behaviour: Exploring the role of social media. *Corporate Social Responsibility and Environmental Management, 25*(6), 1388–1396. https://doi.org/10.1002/csr.1654

Berman, S. L., Wicks, A. C., Kotha, S., & Jones, T. M. (1999). Does stakeholder orientation matter? The relationship between stakeholder management models and firm financial performance. *Academy of Management Journal, 42*(5), 488–506. https://doi.org/10.2307/256972

Birkinshaw, J., Foss, N. J., & Lindenberg, S. M. (2014). Combining purpose with profits. *MIT Sloan Management Review, 55*(3), 49–56.

Blackrock Larry Fink's Letter to CEOs. (2018). *The power of capitalism*. Retrieved July 5, 2024, from https://www.blackrock.com/corporate/investor-relations/larry-fink ceo-letter

Bowen, H. R. (1953). *Social responsibilities of the businessman*. Harper & Row.

Buchholtz, A., & Carroll, A. (2014). *Business and society: Ethics, sustainability, and stakeholder management*. Nelson Education: Cengage Learning.

Bucic, T., Harris, J., & Arli, D. (2012). Ethical consumers among the millennials: A cross-national study. *Journal of Business Ethics, 110*(1), 113–131. https://doi.org/10.1007/s10551-011-1151-z

Carroll, A. B., & Näsi, J. (1997). Understanding stakeholder thinking: Themes from a Finnish conference. *Business Ethics: A European Review, 6*(1), 46–51. https://doi.org/10.1111/1467-8608.00047

Cheah, E.-T., Jamali, D., Johnson, J. E. V., & Sung, M.-C. (2011). Drivers of corporate social responsibility attitudes: The demography of socially responsible investors. *British Journal of Management, 22*(2), 305–323. https://doi.org/10.1111/j.1467-8551.2011.00744.x

Clark, W. H., Jr., & Babson, E. K. (2012). How benefit corporations are redefining the purpose of business corporations. *William Mitchell Law Review, 38*(2), 8.

Clarkson, M. E. (1995). A stakeholder framework for analyzing and evaluating corporate social performance. *Academy of Management Review, 20*(1), 92–117. https://doi.org/10.5465/amr.1995.9503271994

Clarkson, M. (1998). *The corporation and its stakeholders: Classic and contemporary readings*. University of Toronto Press.

Cohen, B., & Muñoz, P. (2017). Entering conscious consumer markets: Toward a new generation of sustainability strategies. *California Management Review, 59*(4), 23–48. https://doi.org/10.1177/0008125617722792

Desai, V. M. (2018). Collaborative stakeholder engagement: An integration between theories of organizational legitimacy and learning. *Academy of Management Journal, 61*(1), 220–244. https://doi.org/10.5465/amj.2016.0315

Donaldson, T., & Preston, L. E. (1995). The stakeholder theory of the corporation: Concepts, evidence, and implications. *Academy of Management Review, 20*(1), 65–91. https://doi.org/10.5465/amr.1995.9503271992

Dorfleitner, G., Halbritter, G., & Nguyen, M. (2015). Measuring the level and risk of corporate responsibility – An empirical comparison of different ESG rating approaches. *Journal of Asset Management, 16*(7), 450–466. https://doi.org/10.1057/jam.2015.31

Eccles, R. G., & Klimenko, S. (2019). The investor revolution. *Harvard Business Review, 97*(3), 106–116.

Elkington, J. (1999). *Cannibals with forks: The triple bottom line of 21st century business*. Capstone.

EU Corporate Sustainability Due Diligence Directive. (2024). Retrieved July 7, 2024, from https://eur-lex.europa.eu/eli/dir/2024/1760/oj

EU Corporate Sustainability Reporting Directive. (2022). Retrieved July 3, 2024, from https://eur-lex.europa.eu/legal-content/EN/TXT/?uri=CELEX:32022L2464

EU Non-Financial Reporting Directive. (2014). Retrieved July 3, 2024, from https://eur-lex.europa.eu/legal-content/EN/TXT/?uri=celex%3A32014L0095

EU Taxonomy for Sustainable Activities. (2020). Retrieved July 3, 2024, from https://eur-lex.europa.eu/legal-content/EN/TXT/?uri=CELEX:32020R0852

European Commission. (2020). *The European Green Deal*. Retrieved July 3, 2024, from https://commission.europa.eu/strategy-and-policy/priorities-2019-2024/european-green-deal_en

Financial Times. (2023, July 4). *Rating the ESG rating agencies*. Retrieved July 5, 2024, from https://www.ft.com/content/e9eaa11a-31e0-4f60-9a65-b6883546e8da

Freeman, R. E. (1984). *Strategic management: A stakeholder approach*. Pitman.

Freeman, R. E., Wicks, A. C., & Parmar, B. (2004). Stakeholder theory and "the corporate objective revisited". *Organization Science, 15*(3), 364–369. https://doi.org/10.1287/orsc.1040.0066

Freudenreich, B., Lüdeke-Freund, F., & Schaltegger, S. (2020). A stakeholder theory perspective on business models: Value creation for sustainability. *Journal of Business Ethics, 166*(1), 3–18. https://doi.org/10.1007/s10551-019-04112-z

Friedman, M. (1970, September 13). The social responsibility of business is to increase its profits. *New York Times*.

Gehman, J., Grimes, M. G., & Cao, K. (2019). Why we care about certified B corporations: From valuing growth to certifying values practices. *Academy of Management Discoveries, 5*(1), 97–101. https://doi.org/10.5465/amd.2018.0074

George, G., Haas, M. R., McGahan, A. M., Schillebeeckx, S. J. D., & Tracey, P. (2023). Purpose in the for-profit firm: A review and framework for management research. *Journal of Management, 49*(6), 1841–1869. https://doi.org/10.1177/01492063211006450

Global Reporting Initiative. (1997). *The global leader for impact reporting*. Retrieved July 2, 2004, from https://www.globalreporting.org/

Harrison, J. S., & Wicks, A. C. (2013). Stakeholder theory, value, and firm performance. *Business Ethics Quarterly, 23*(1), 97–124. https://doi.org/10.5840/beq20132314

Hiller, J. S. (2013). The benefit corporation and corporate social responsibility. *Journal of Business Ethics, 118*(2), 287–301. https://doi.org/10.1007/s10551-012-1580-3

Hillman, A. J., & Keim, G. D. (2001). Shareholder value, stakeholder management, and social issues: What's the bottom line? *Strategic Management Journal, 22*(2), 125–139. https://doi.org/10.1002/1097-0266(200101)22:2<125::aid-smj150>3.0.co;2-h

References

Jensen, M. C. (2001). Value maximization, stakeholder theory, and the corporate objective function. *Journal of Applied Corporate Finance, 14*(3), 8–21. https://doi.org/10.1111/j.1745-6622.2001.tb00434.x

Jones, T. M. (1995). Instrumental stakeholder theory: A synthesis of ethics and economics. *Academy of Management Review, 20*(2), 404–437. https://doi.org/10.5465/amr.1995.9507312924

Kim, S., Karlesky, M. J., Myers, C. G., & Schifeling, T. (2016). Why companies are becoming B corporations. *Harvard Business Review, 17*, 1–5.

Levillain, K., & Segrestin, B. (2019). From primacy to purpose commitment: How emerging profit-with-purpose corporations open new corporate governance avenues. *European Management Journal, 37*(5), 637–647. https://doi.org/10.1016/j.emj.2019.07.002

Marquis, C. (2020). *Better business: How the B Corp movement is remaking capitalism*. Yale University Press.

Martini, A. (2021). Socially responsible investing: From the ethical origins to the sustainable development framework of the European Union. *Environment, Development and Sustainability, 23*(11), 16874–16890. https://doi.org/10.1007/s10668-021-01375-3

McGuire, J. B., Sundgren, A., & Schneeweis, T. (1988). Corporate social responsibility and firm financial performance. *Academy of Management Journal, 31*(4), 854–872. https://doi.org/10.5465/256342

Mitchell, R. K., Agle, B. R., & Wood, D. J. (1997). Toward a theory of stakeholder identification and salience: Defining the principle of who and what really counts. *Academy of Management Review, 22*(4), 853–886. https://doi.org/10.5465/amr.1997.9711022105

Mitchell, J. R., Mitchell, R. K., Hunt, R. A., Townsend, D. M., & Lee, J. H. (2022). Stakeholder engagement, knowledge problems and ethical challenges. *Journal of Business Ethics, 175*(1), 75–94. https://doi.org/10.1007/s10551-020-04550-0

Morsing, M., & Schultz, M. (2006). Corporate social responsibility communication: Stakeholder information, response and involvement strategies. *Business Ethics: A European Review, 15*(4), 323–338. https://doi.org/10.1111/j.1467-8608.2006.00460.x

Sen, S., Bhattacharya, C. B., & Korschun, D. (2006). The role of corporate social responsibility in strengthening multiple stakeholder relationships: A field experiment. *Journal of the Academy of Marketing Science, 34*(2), 158–166. https://doi.org/10.1177/0092070305284978

United Nations Environment Programme Finance Initiative Report. (2004). *Who cares wins – connecting financial markets to a changing world*. Retrieved July 1, 2024, from https://www.unepfi.org/fileadmin/events/2004/stocks/who_cares_wins_global_compact_2004.pdf

United Nations Global Compact. (2000). Retrieved July 2, 2004, from https://unglobalcompact.org/

United Nations Millennium Development Goals. (2000). Retrieved July 1, 2024, from https://www.un.org/millenniumgoals/

United Nations Principles for Responsible Investment. (2006). Retrieved July 2, 2024, from https://www.unpri.org/

United Nations Sustainable Development Goals. (2015). Retrieved July 1, 2024, from https://sdgs.un.org/goals

Wolf, J. (2013). Improving the sustainable development of firms: The role of employees. *Business Strategy and the Environment, 22*(2), 92–108. https://doi.org/10.1002/bse.1731

Yu, E. P. Y., Van Luu, B., & Chen, C. H. (2020). Greenwashing in environmental, social and governance disclosures. *Research in International Business and Finance, 52*, 101192. https://doi.org/10.1016/j.ribaf.2020.101192

Chapter 3
Governance and Sustainability Practices in Family Businesses

3.1 Relationship Between Governance and Sustainability

Over the past two decades, the literature on corporate governance and sustainability has grown significantly. This growth reflects a rising scholarly interest in exploring how and to what extent corporate governance practices can enhance sustainability outcomes (e.g., Aguilera et al., 2021; Hussain et al., 2018; Michelon & Parbonetti, 2012; Naciti et al., 2022).

As demonstrated by an extensive body of research, corporate governance plays a pivotal role in influencing a company's sustainability efforts (Aras & Crowther, 2008). In fact, it can ensure that sustainability is a core consideration in the decision-making, risk management, reporting, stakeholder engagement, and ethical conduct of companies. A robust corporate governance framework that addresses sustainability issues enables companies to recognize and meet the interests of both shareholders and stakeholders. This approach not only responds to the increasing demand for tangible sustainability actions (Barnett et al., 2018) but also contributes to long-term corporate success and generates positive societal impact.

The literature on corporate governance and sustainability is characterized by considerable diversity and heterogeneity, with differing findings depending on the topics studied and the theoretical frameworks used. Research has focused primarily on the ways in which factors such as board composition (Harjoto et al., 2019; Ricart et al., 2005; Webb, 2004; Zhang, 2012), ownership structure (Arduino et al., 2024; Rees & Rodionova, 2015; Semenova & Hassel, 2019), and executive compensation (Haque, 2017; Hong et al., 2016; Mahoney & Thorn, 2006) impact sustainability performance (Lozano & Martínez-Ferrero, 2022; Pozzoli et al., 2022; Shaukat et al., 2016) and sustainability disclosure and reporting (Amran et al., 2014; Husted & de Sousa-Filho, 2019; Suttipun, 2021).

The relationship between governance and sustainability has increasingly attracted interest outside academia as well. Many leading institutions (e.g., the OECD and the

EU) have offered guidelines for promoting sustainable governance and conducting analyses and research on the adoption of good corporate governance and responsible practices among companies.

According to a recent study conducted by the OECD (2023), at the end of 2022, in 24% of 49 jurisdictions analyzed, sustainability-related disclosure is recommended by codes or principles, often following a "comply or explain"[1] approach set by regulators or stock exchanges. Regarding applicability for sustainability-related disclosure, 25 jurisdictions require or recommend that only listed companies disclose sustainability information, whereas the remaining 24 jurisdictions extend these disclosure requirements to both listed and nonlisted companies. In the EU, the 2014 NFRD was the primary source for sustainability disclosure requirements for several years, mandating that listed companies and certain nonlisted public interest entities disclose sustainability information. Nevertheless, companies had the flexibility to choose their disclosure standards. However, the 2022 CSRD introduced significant changes, requiring companies to follow the new EU Sustainability Reporting Standards (ESRS) developed by the European Financial Reporting Advisory Group (EFRAG). Europe also leads in the proportion of companies that obtain assurance for their sustainability-related information; in fact, 81% of companies disclosing such information engage an independent third party.

With respect to the board of directors, in half of the jurisdictions analyzed by the OECD, boards are explicitly required (by the law or regulations or listing rules) or recommended (by the codes) to approve sustainability-related policies, such as sustainability plans, targets, and internal control policies, for managing sustainability risk. Globally, companies that account for half of the total market capitalization have a board committee dedicated to sustainability. Boards can indeed either create a new committee or extend the responsibilities of an existing committee to enhance their oversight of sustainability-related governance practices, disclosures, strategies, and risk management.

In terms of board gender diversity, countries have taken significant steps in recent years to promote greater female representation on boards by adopting various approaches, such as implementing mandatory quotas or setting recommended targets.

The growing focus on shareholder remuneration has both benefited from and driven the enhancement of disclosure requirements. In fact, all countries mandate or recommend that companies disclose their remuneration policies, and nearly all require or recommend the disclosure of total aggregate remuneration. With respect to executive pay, boards of directors can integrate sustainability-related metrics into their remuneration decisions. Sustainability-linked executive compensation has become widespread, especially among major publicly listed companies in Europe and the United States. In Europe, over 70% of companies by market capitalization

---

[1] The "comply or explain" approach is a regulatory mechanism that allows companies to either align with established practices or openly communicate their reasons for their noncompliance with them.

link executive compensation to sustainability factors, followed by the United States, where between 50% and 60% of companies do the same.

With the aim of increasing trust in their long-term business strategies, companies can implement policies that promote shareholder engagement. According to the Global Corporate Sustainability Report 2024 by the OECD (2024), as of 2022, 81% of companies globally by market capitalization disclose policies for shareholder engagement, detailing mechanisms for shareholders to question the board, interact with management, or submit proposals at shareholder meetings. The highest rates of such policies are seen in the United States, where 93% of market capitalization is covered, and in Europe, where 85% of market capitalization is included.

Furthermore, 70% of companies by market capitalization provide details on their stakeholder engagement practices and the ways in which they involve stakeholders in decision-making processes. In Europe, this percentage is notably higher, with nearly 90% of companies disclosing such information. To enhance stakeholder engagement, companies may implement various mechanisms, such as including employee representatives on the board or establishing workers' councils that incorporate employee perspectives in key decisions.

3.2 Understanding Sustainability in the Family Business Context

According to the Brundtland Report, officially titled "Our Common Future," published in 1987 by the UN's World Commission on Environment and Development (WCED), sustainable development can be defined as "development that meets the needs of the present without compromising the ability of future generations to meet their own needs" (Brundtland, 1987).

The drive toward sustainability is particularly significant in family firms. Family businesses inherently have a long-term vision, with the transition to future generations as a central focus and the concept of legacy as a fundamental imperative for the family (Hammond et al., 2016; Le Breton-Miller & Miller, 2006).

The SEW perspective (Gómez-Mejía et al., 2007) suggests that family firms prioritize nonfinancial goals or "affective endowments" that fulfill the family's emotional needs (Berrone et al., 2012). These firms often seek to enhance their reputation, be seen as responsible agents, and gain social legitimacy (Crossley et al., 2021; Zellweger et al., 2013).

In the extended SEW framework (Miller & Le Breton-Miller, 2014), reputation is viewed as a core socioemotional asset in family firms that is tied to identity, legacy, and long-term orientation. It shapes strategic decisions, as preserving family name and trust is often prioritized over short-term gains.

SEW theory also elucidates stakeholder relationships, revealing how family firms prioritize and cultivate strong connections with their employees, customers, and community because of their emphasis on social ties and proactive stakeholder engagement. These strong relationships can enhance a firm's reputation, as stakeholders often appreciate the family firm's commitment to long-term relationships and community-oriented values (Cennamo et al., 2012).

In line with this focus on strong social connections, the "Binding Social Ties" dimension of the FIBER model (Berrone et al., 2012) further highlights how emotional and relational connections within family firms drive sustainability initiatives. These ties promote mutual trust and long-term commitment, motivating family firms to prioritize sustainability over short-term profits. This dedication to long-term sustainability goals extends beyond the immediate family, influencing interactions with external stakeholders and reinforcing the business' broader sustainability efforts.

Many studies document that family firms exhibit superior sustainability performance compared with their nonfamily counterparts (e.g., Berrone et al., 2010; Dyer & Whetten, 2006; García-Sánchez et al., 2020). However, some scholars have argued that family firms may not always be more socially responsible, as family members might prioritize their personal interests over broader social concerns, owing to the "dark side" of SEW (Kellermanns et al., 2012; Morck & Yeung, 2004).

Cruz et al. (2014) argued that while family firms respond more to external stakeholders, they may limit social actions affecting internal stakeholders to retain control and preserve SEW. For example, they might hire family members despite their lack of qualifications (Chua et al., 2009) or separate family members' compensation from actual performance results (Cruz et al., 2010). This raises questions about whether family firms adjust their governance to meet sustainability demands, as they might resist changes that could lessen their influence. Therefore, scholars debate whether the governance adjustments of family firms are merely related to compliance with legal requirements or if they lead to significant changes in corporate strategies that address stakeholder expectations (Carroll & Shabana, 2010).

According to prior research, corporate sustainability strategies range from reactive to proactive. Firms that adopt the reactive approach address external pressures as they occur, often using symbolic measures to manage their public image and demonstrate compliance with social and legal obligations. Conversely, firms that take a proactive stance develop voluntary strategies that go beyond mere regulatory compliance and implement meaningful actions to meet stakeholder expectations (Haji & Anifowose, 2016; Sharma & Sharma, 2011).

A recent study by Arduino et al. (2024) revealed that family firms tend to have lower ESG transparency than nonfamily firms do. This tendency is driven by a desire to maintain privacy and control, as disclosing extensive ESG information could lead to increased scrutiny. However, the presence of institutional investors moderates this relationship, playing a crucial role in promoting greater openness and transparency in ESG practices.

Delmas and Gergaud (2014) argued that when family firms adopt sustainability certifications, their ability to signal sustainability to stakeholders is enhanced, which helps mitigate concerns about greenwashing (Delmas & Burbano, 2011). For example, the B Corp certification provides notable recognition for companies that effectively balance profit with purpose and show a strong commitment to positive social and environmental impacts. This certification has received increasing attention from scholars in recent years (e.g., Cao et al., 2017; Honeyman & Jana, 2019). Furthermore, interest in hybrid business models that emphasize nonfinancial dimensions and values close to those seen in family firms has increased (Grassl, 2011).

3.3 How to Foster Sustainable Governance

Fostering sustainable governance involves implementing practices and policies that ensure that a company not only complies with current regulations but also actively promotes long-term environmental, social, and economic interests. This includes integrating sustainability into the company's governance framework by embedding sustainability objectives within the overall strategy and decision-making processes, as well as establishing clear policies to address sustainability issues and set specific goals for reducing environmental and social impacts.

To advance sustainable governance and increase attention to stakeholder interests, various regulatory measures—both at the supranational and national levels—as well as guidelines and best practices—have been introduced in recent years.

Sustainability reporting standards and frameworks, such as the Global Reporting Initiative (GRI) (Global Reporting Initiative, 1997) and the Sustainability Accounting Standards Board (SASB) (Sustainability Accounting Standards Board, 2011), have been developed to promote transparent and consistent information regarding the sustainability practices and performances of companies.

Board oversight and accountability have been enhanced, often with a clear definition of the board's role in sustainability. For example, the UK's Companies Act of 2006 (Legislation.gov.uk, 2006) requires directors to consider the interests of employees, suppliers, consumers, and other stakeholders, as well as the community and the environment. Similarly, France's Loi Pacte in 2019 (Legifrance, 2019) redefined companies to include not only the interests of shareholders but also the obligation to account for the social and environmental dimensions in company strategies.

In Italy, one of the most significant innovations of the Corporate Governance Code issued in 2020 by Borsa Italiana is the establishment of sustainable success as "the objective that guides the actions of the board of directors, materializing in the creation of long-term value for the benefit of shareholders, taking into account the interests of other stakeholders relevant to the company" (Borsa Italiana, 2020). Thus, the pursuit of sustainable success becomes a fundamental responsibility that guides a company's management and is divided into two key areas:

- The creation of long-term value for the benefit of shareholders: this primary objective extends beyond the traditional shareholder value creation paradigm. Although profit cannot be limited to short-term views, it remains the ultimate goal for shareholders and, consequently, a key responsibility for directors. Thus, the board of directors needs to carefully assess policies when elaborating strategic plans and consider these factors in management decisions, especially significant ones.

– Consideration of the interests of other stakeholders relevant to the company: the interests of the various stakeholders who impact or are impacted by the company must be recognized and prioritized through a systematic approach to identify which stakeholders' interests should take precedence. After these priorities are established, companies should conduct a detailed mapping of each stakeholder's concerns and objectives. This mapping can help companies define clear goals and strategies for engaging with different stakeholder groups to ensure that their interests are adequately addressed in the company's decision-making processes (Walker et al., 2008).

Additionally, as the Code states that "The board of directors promotes, in the most appropriate forms, dialogue with shareholders and other stakeholders relevant to the company," companies must develop effective communication strategies tailored to engage the different stakeholder groups. This implies that companies need to create appropriate channels and methods to engage in meaningful interaction and feedback, thereby strengthening their relationships with various stakeholders and addressing stakeholders' concerns effectively.

According to Kujala et al. (2022, p. 4), *"Stakeholder engagement refers to the aims, activities, and impacts of stakeholder relations in a moral, strategic, and/or pragmatic manner."* Indeed, active stakeholder engagement is becoming increasingly important for companies. In fact, they are expected to develop engagement mechanisms by establishing channels for meaningful dialog with their internal and external stakeholders, including employees, customers, suppliers, and communities. Companies should then incorporate stakeholder feedback into their decision-making processes and sustainability strategies (Noland & Phillips, 2010).

In terms of executive compensation, linking pay to sustainability goals is crucial for companies. When sustainability metrics are incorporated into compensation packages, leaders' incentives are aligned with long-term sustainability objectives (Adu et al., 2022; Velte, 2024). Companies can include either quantitative or qualitative goals. Although some targets can be difficult to measure, investors are increasingly requesting that the sustainability goals for which managers are held accountable are clear, specific, and measurable. Many companies have increasingly included nonfinancial metrics such as ESG metrics in their incentive plans to strengthen their commitment to sustainability goals and guarantee the accountability of managers for advancing in the environmental, social, and governance dimensions.

Furthermore, the introduction of new legal forms, such as benefit corporations, which allow companies to include the pursuit of public benefits in their articles of incorporation, represents a novel approach to integrating sustainability into a company's governance. These legal structures align legal obligations with broader societal and environmental goals, therefore promoting a long-term positive impact. Starting in the U.S. in 2010, many jurisdictions have introduced the benefit corporation legal form, and an increasing number of companies have adopted it, implementing a stakeholder-centric business model that pursues the public benefit.

The promotion of sustainable governance can have significant implications for family businesses, affecting both their operations and their broader societal impact.

Table 3.1 Fostering sustainable governance: Institutional and firm-level initiatives

Level and Initiatives
Institutional-level
Setting laws, regulations and guidelines
Defining disclosure and reporting standards
Introducing new legal forms formally committed to pursuing public benefit purposes alongside profit goals
Firm-level
Embedding sustainability into strategy and culture
Fostering multistakeholder engagement
Enhancing board oversight and accountability over sustainability
Linking executive pay to sustainability-related goals

Source: Author's elaboration

Integrating a sustainability strategy into the business is crucial. This implies that family firms are reshaping the organizational culture and the business model to incorporate societal and environmental goals, aligning with the SEW tenets that advocate for the socio-environmental commitment of family businesses (Berrone et al., 2010). However, the implementation of CSR practices necessitates investments that could increase a company's risk levels, potentially jeopardizing its long-term sustainability (Dal Maso et al., 2020).

In terms of governance, to promote socially responsible behavior, family firms should have a substantial family presence within both the management team and the board of directors. In fact, prior research has indicated that such representation is linked to improved sustainability outcomes (López-González et al., 2019). On the other hand, a strong family presence appears to inhibit a company's engagement in communicating sustainability efforts because of the negative impact of family involvement in management on the disclosure of CSR information (Venturelli et al., 2021). Table 3.1 summarizes the main institutional and firm-level initiatives that can be implemented to foster sustainable governance.

3.4 Governance and Sustainability in Italy

In Italy, the corporate governance system adheres to the Latin model, which is characterized by its distinct approach to corporate governance compared with other models, such as the Anglo-Saxon or the German models (Melis, 2000). However, in recent years, the Italian governance model has increasingly aligned with the Anglo–Saxon model, a practice driven by external pressures to align with international best practices (Zattoni, 2019).

The Latin model, which is prevalent in several Southern European countries, has specific features and characteristics that shape the management and oversight of companies.

Examining the institutional framework and stock market development, Italy appears to have relatively low legal protection for investors and weak legal enforcement, as well as underdeveloped capital markets (Aganin & Volpin, 2005). As a result, minority shareholders demand a greater return on their investments to compensate for the elevated risk of expropriation by management or controlling shareholders (La Porta et al., 1998).

In terms of ownership structure, Italian nonfinancial listed firms are characterized by a high level of concentration, typically with a predominant main shareholder or blockholder. This main shareholder often holds a significant part of the company's shares, exerting substantial influence over corporate decisions, particularly in the context of pyramidal ownership structures[2] (Bianchi et al., 1997; Melis, 2000). The ownership of Italian listed firms is primarily concentrated in the hands of families or the State, with limited institutional investor involvement (CONSOB, 2023).[3]

According to a study conducted by the OECD (2023), Italy is the tenth country (out of 49) in terms of ownership concentration, as the largest and three largest shareholder(s) hold more than 50% of the company's equity capital.

Italian board structures are governed by specific laws and regulations that outline various organizational forms and governance mechanisms. The main governance structures allowed by law in Italy are as follows:

- The traditional model (*modello tradizionale*): in this model, both the board of directors and the board of statutory auditors (*collegio sindacale*) are selected by the shareholders' meeting. The board of directors can delegate daily management responsibilities to one or more executive directors or to an executive committee. According to CONSOB (2023), at the end of 2022, the traditional model remains mostly adopted by Italian companies with ordinary shares listed on Euronext Milan.
- The two-tier model (*modello dualistico*): in this model, the supervisory board is appointed by the shareholders' meeting, whereas the management board is designated by the supervisory board, unless the bylaws specify that the shareholders' meeting is responsible for this appointment. The supervisory board does not have operational or executive powers, but the bylaws can assign higher-level management responsibilities.

---

[2] In pyramidal groups, a holding company exerts control over the majority of voting rights in the subsidiary companies within the group, either directly or indirectly. Ultimate control of the group typically is in the hands of an individual entrepreneur, a family, or a coalition. This structure is often employed to maximize the leverage of resources controlled relative to the capital invested, thus ensuring that control is maintained. The organizational structures of these groups can be quite complex, making it challenging to reproduce their control structures precisely, especially on an international scale, despite ownership disclosure regulations (Melis, 2000).

[3] According to CONSOB (2023), in 2023, institutional investors were present in 51 companies, a decrease from 67 in 2019. This decline is particularly significant among foreign institutional investors, who held major stakes in 40 companies, down from 55 in 2019. Conversely, Italian institutional investors have seen a slight and steady increase, with significant holdings in 17 companies by the end of 2023, compared with 14 in 2019.

3.4 Governance and Sustainability in Italy

- The one-tier model (*modello monistico*): in this model, the shareholders' meeting appoints the board of directors, whereas the management control committee consists of nonexecutive, independent board members. The board of directors can delegate daily management responsibilities to one or more managing directors or to an executive committee.

The representation of minority shareholders on the board is regulated by several legal provisions to ensure that minority interests are considered in corporate governance (e.g., the Italian Civil Code and the Corporate Governance Code). For example, at least one board member must be elected from the slate of candidates presented by shareholders that own a minimum threshold of the company's share capital. Shareholder agreements can sometimes provide additional rights for minority shareholders, such as specific provisions for board representation, although these rights are subject to the company's articles of association.

With respect to board independence, provisions can vary according to a company's ownership structure. For companies within a pyramidal group controlled by a listed parent company, the listing regulations require that the board of directors is composed of a majority of independent directors, who must not be members of the parent company's board. For large companies without controlling shareholders, the Borsa Italiana's Corporate Governance Code recommends that the majority of directors should be independent. In contrast, for large companies with controlling shareholders, the Code recommends that at least one-third of the directors should be independent. The goal of this recommendation is to increase the objectivity and effectiveness of the oversight functions of the board of directors, particularly in companies with concentrated ownership.

With respect to CEO duality, the Corporate Governance Code does not explicitly mandate the separation of the roles of chairperson and CEO. However, it does address CEO duality by requiring that if the same individual holds both positions, a lead independent director (LID) must be appointed to the board. This lead independent director plays a crucial role in providing oversight and guaranteeing that the activities of the board are carried out independently, therefore, helping to mitigate potential conflicts of interest and increasing the board's overall effectiveness. The appointment of a lead independent director also helps sustain the principles of balance and accountability within the board, even when the roles of the chairperson and CEO are consolidated (Krause et al., 2017).

In relation to female board representation, the introduction of the Golfo–Mosca Law (Law No. 120/2011) (Gazzetta Ufficiale della Repubblica Italiana, 2011) represented an important advancement by requiring that at least one-third of the board members of Italian publicly listed companies and state-owned enterprises be women. Since this law was enacted, the presence of female directors on Italian boards has increased steadily. The subsequent Law 160/2019 (Gazzetta Ufficiale della Repubblica Italiana, 2019) further advanced board gender diversity by raising the quota from 33% to 40%, starting in 2020, and extending its application to cover 6 additional board appointments over nearly 18 years. According to the OECD (2023), at the end of 2022, Italy was fourth out of 49 countries in terms of board gender diversity, surpassing the European average.

In Italy, executive compensation is regulated by a combination of statutory provisions, regulations from regulatory bodies such as CONSOB (the Italian Companies and Stock Exchange Commission), and principles outlined in the Corporate Governance Code. According to the Code, companies must provide a comprehensive disclosure of their executive compensation policies and practices in their corporate governance reports, also indicating the performance metrics utilized. Additionally, the Code recommends that listed companies establish a remuneration committee that should primarily consist of independent directors. This committee is responsible for determining and reviewing the compensation for the CEO and for other executives to ensure that compensation packages are aligned with the company's long-term goals and performance. The Code also promotes the integration of sustainability targets into executive compensation and encourages companies to include ESG criteria in their performance metrics. Indeed, companies are increasingly incorporating social and environmental targets into their executive compensation plans. Integrating explicit sustainability-linked objectives into these plans increases CEOs' commitment to sustainability issues (Berrone & Gomez-Mejia, 2009). Moreover, a recent study by Almici (2023) indicates that Italian firms that embed sustainability into executive remuneration policies tend to achieve better nonfinancial performance.

Regarding the integration of sustainability considerations into corporate governance, the Code recommends that the board of directors oversee the company's sustainability practices and guarantee that ESG factors are considered in the company's strategies. Although the Code does not explicitly require the creation of a dedicated sustainability committee, many Italian companies choose to establish such committees. These sustainability (or CSR or ESG) committees are responsible for advising the board on sustainability strategies, ensuring compliance with regulations, and embedding sustainability into the broader corporate strategy (Orazalin, 2020). Alternatively, some Italian companies integrate sustainability oversight into existing board committees, such as audit or risk committees. When this approach is adopted, sustainability issues are considered along with financial and risk management aspects.

Given the evolving nature of sustainability regulations, particularly in Europe, and changes in corporate governance practices, companies are increasingly expected to adapt and enhance their governance structures to address ESG considerations effectively.

For example, according to a study on the engagement policies of Italian listed companies conducted by CONSOB (2022), many Italian firms adhere to the Corporate Governance Code's recommendation for boards to establish policies for managing dialog with shareholders on ESG issues.

Furthermore, since its emergence in Italy in 2013, the number of firms applying for B Corp certification has progressively increased. As of the end of June 2024, there were 288 certified B-Corps in Italy (B Corporation, 2024). Among them, 30 are family firms: 3 are large, 5 are medium-sized, and the remaining are small. In Italy, all certified B-Corps are also required to become Benefit Corporations, therefore, changing their legal status to formally integrate a purpose beyond profit into their statutes.

3.4 Governance and Sustainability in Italy

Focusing on Italian family firms, the Association of Italian Family Businesses (AIDAF) in collaboration with NATIVA (the first Italian and European company to become a Benefit Corporation) developed the AIDAF Legacy Book (NATIVA, 2024), a document that provides insights into their approach to sustainability and illustrates the commitments and challenges faced by family businesses in creating a positive and lasting legacy for society and the planet.

Italian family firms indeed face many sustainability challenges: seeking to integrate various aspects of sustainability into their strategies, they strive to balance business goals with social and environmental impact.

From an environmental perspective, family firms are expected to adopt more sustainable practices to reduce their ecological footprint under the pressure of more stringent regulations. This includes, for example, mitigating climate change by reducing greenhouse gas emissions, therefore pursuing a more sustainable business model.

Considering the social dimension, family firms are called to address societal concerns by enacting initiatives to respond to both internal and external stakeholders' needs, such as investing in the welfare of their employees and contributing to the local community where they operate.

Increasing EU sustainability regulations pose significant operational challenges for family firms, especially in terms of compliance costs, administrative burden, and the need to adapt internal structures and processes. Smaller family firms are particularly exposed, as they often lack the financial resources and the managerial capacity necessary to adopt environmentally responsible technologies or implement sustainable practices effectively. Expenses such as upgrading equipment and machinery and monitoring environmental impacts may represent a substantial investment for these businesses. Moreover, new reporting and transparency requirements can place additional pressure on family firms' management and governance structures. Many family firms operate with streamlined administrative systems and informal governance structures, which may conflict with the increasingly structured and data-driven compliance demands. These external pressures may force family businesses to adopt more formal governance and sustainability reporting systems, which could challenge their internal dynamics. Although this transition might bring long-term benefits such as increased legitimacy and greater stakeholder trust, it can also be a complex and resource-intensive process for family firms in the short term.

References

Adu, D. A., Flynn, A., & Grey, C. (2022). Executive compensation and sustainable business practices: The moderating role of sustainability-based compensation. *Business Strategy and the Environment, 31*(3), 698–736. https://doi.org/10.1002/bse.2913

Aganin, A., & Volpin, P. (2005). The history of corporate ownership in Italy. In R. K. Morck (Ed.), *A history of corporate governance around the world: Family business groups to professional managers* (pp. 325–366). University of Chicago Press.

Aguilera, R. V., Aragón-Correa, J. A., Marano, V., & Tashman, P. A. (2021). The corporate governance of environmental sustainability: A review and proposal for more integrated research. *Journal of Management, 47*(6), 1468–1497. https://doi.org/10.1177/0149206321991212

Almici, A. (2023). Does sustainability in executive remuneration matter? The moderating effect of Italian firms' corporate governance characteristics. *Meditari Accountancy Research, 31*(7), 49–87. https://doi.org/10.1108/medar-05-2022-1694

Amran, A., Lee, S. P., & Devi, S. S. (2014). The influence of governance structure and strategic corporate social responsibility toward sustainability reporting quality. *Business Strategy and the Environment, 23*(4), 217–235. https://doi.org/10.1002/bse.1767

Aras, G., & Crowther, D. (2008). Governance and sustainability: An investigation into the relationship between corporate governance and corporate sustainability. *Management Decision, 46*(3), 433–448. https://doi.org/10.1108/00251740810863870

Arduino, F. R., Buchetti, B., & Harasheh, M. (2024). The veil of secrecy: Family firms' approach to ESG transparency and the role of institutional investors. *Finance Research Letters, 62*, 105243. https://doi.org/10.1016/j.frl.2024.105243

B Corporation. (2024). *B Lab Italy*. Retrieved July 6, 2024, from https://bcorporation.eu/country_partner/italy/

Barnett, M. L., Henriques, I., & Husted, B. W. (2018). Governing the void between stakeholder management and sustainability. In S. Dorobantu, R. V. Aguilera, J. Luo, & F. J. Milliken (Eds.), *Sustainability, stakeholder governance, and corporate social responsibility* (pp. 121–143). Emerald Publishing Limited.

Berrone, P., Cruz, C., & Gomez-Mejia, L. R. (2012). Socioemotional wealth in family firms: Theoretical dimensions, assessments approaches and agenda for future research. *Family Business Review, 25*(3), 258–279. https://doi.org/10.1177/0894486511435355

Berrone, P., Cruz, C., Gomez-Mejia, L. R., & Larraza-Kintana, M. (2010). Socioemotional wealth and corporate responses to institutional pressures: Do family-controlled firms pollute less? *Administrative Science Quarterly, 55*(1), 82–113. https://doi.org/10.2189/asqu.2010.55.1.82

Berrone, P., & Gomez-Mejia, L. R. (2009). Environmental performance and executive compensation: An integrated agency-institutional perspective. *Academy of Management Journal, 52*(1), 103–126. https://doi.org/10.5465/amj.2009.36461950

Bianchi, M., Bianco, M., & Enriques, L. (1997). *Ownership, pyramidal groups and separation between ownership and control in Italy*. European Corporate Governance Network.

Borsa Italiana. (2020). *Corporate governance code*. Retrieved July 8, 2024, from https://www.borsaitaliana.it/comitato-corporate-governance/codice/2020.pdf

Brundtland, G. (1987). *Report of the world commission on environment and development: Our common future*. United Nations General Assembly document A/42/427.

Cao, K., Gehman, J., & Grimes, M. G. (2017). Standing out and fiting in: Charting the emergence of certified B corporations by industry and region. In A. C. Corbett & J. A. Katz (Eds.), *Hybrid ventures: Advances in entrepreneurship, firm emergence and growth* (Vol. 19, pp. 1–38). Emerald Publishing Limited.

Carroll, A. B., & Shabana, K. M. (2010). The business case for corporate social responsibility: A review of concepts, research and practice. *International Journal of Management Reviews, 12*(1), 85–105. https://doi.org/10.1111/j.1468-2370.2009.00275.x

Cennamo, C., Berrone, P., Cruz, C., & Gomez-Mejia, L. R. (2012). Socioemotional wealth and proactive stakeholder engagement: Why family-controlled firms care more about their stakeholders. *Entrepreneurship Theory and Practice, 36*(6), 1153–1173. https://doi.org/10.1111/j.1540-6520.2012.00543.x

Chua, J. H., Chrisman, J. J., & Bergiel, E. B. (2009). An agency theoretic analysis of the professionalized family firm. *Entrepreneurship Theory and Practice, 33*(2), 355–372. https://doi.org/10.1111/j.1540-6520.2009.00294.x

CONSOB. (2022). *Report on corporate governance of Italian listed companies with the Addendum Engagement*. Retrieved July 12, 2024, from https://www.consob.it/documents/11973/545079/rcg2022.pdf/33d25582-ade3-d06b-7395-654be6cd7e43?t=1682665906755

References

CONSOB. (2023). *Report on corporate governance of Italian listed companies*. Retrieved July 12, 2024, from https://www.consob.it/documents/11973/545079/rcg2023.pdf/a0a36447-9dba-8a5d-c79a-d0499fdffbd8

Crossley, R. M., Elmagrhi, M. H., & Ntim, C. G. (2021). Sustainability and legitimacy theory: The case of sustainable social and environmental practices of small and medium-sized enterprises. *Business Strategy and the Environment, 30*(8), 3740–3762. https://doi.org/10.1002/bse.2837

Cruz, C. C., Gómez-Mejia, L. R., & Becerra, M. (2010). Perceptions of benevolence and the design of agency contracts: CEO-TMT relationships in family firms. *Academy of Management Journal, 53*(1), 69–89. https://doi.org/10.5465/amj.2010.48036975

Cruz, C., Larraza-Kintana, M., Garcés-Galdeano, L., & Berrone, P. (2014). Are family firms really more socially responsible? *Entrepreneurship Theory and Practice, 38*(6), 1295–1316. https://doi.org/10.1111/etap.12125

Dal Maso, L., Basco, R., Bassetti, T., & Lattanzi, N. (2020). Family ownership and environmental performance: The mediation effect of human resource practices. *Business Strategy and the Environment, 29*(3), 1548–1562. https://doi.org/10.1002/bse.2452

Delmas, M. A., & Burbano, V. C. (2011). The drivers of greenwashing. *California Management Review, 54*(1), 64–87. https://doi.org/10.1525/cmr.2011.54.1.64

Delmas, M. A., & Gergaud, O. (2014). Sustainable certification for future generations. *Family Business Review, 27*(3), 228–243. https://doi.org/10.1177/0894486514538651

Dyer, W. G., & Whetten, D. A. (2006). Family firms and social responsibility: Preliminary evidence from the S&P 500. *Entrepreneurship Theory and Practice, 30*(6), 785–802. https://doi.org/10.1111/j.1540-6520.2006.00151.x

García-Sánchez, I. M., Martín-Moreno, J., Khan, S. A., & Hussain, N. (2020). Socio-emotional wealth and corporate responses to environmental hostility: Are family firms more stakeholder oriented? *Business Strategy and the Environment, 30*(2), 1003–1018. https://doi.org/10.1002/bse.2666

Gazzetta Ufficiale della Repubblica Italiana. (2011). *Italian Law No. 120/2011*. Retrieved July 13, 2024, from https://www.gazzettaufficiale.it/eli/id/2011/07/28/011G0161/sg

Gazzetta Ufficiale della Repubblica Italiana. (2019). *Italian Law No. 160/2019*. Retrieved July 13, 2024, from https://www.gazzettaufficiale.it/eli/id/2019/12/30/19G00165/sg

Global Reporting Initiative. (1997). *The global leader for impact reporting*. Retrieved July 2, 2004, from https://www.globalreporting.org/

Gómez-Mejía, L. R., Haynes, K. T., Núñez-Nickel, M., Jacobson, K. J. L., & Moyano-Fuentes, J. (2007). Socioemotional wealth and business risks in family-controlled firms: Evidence from Spanish olive oil mills. *Administrative Science Quarterly, 52*(1), 106–137. https://doi.org/10.2189/asqu.52.1.106

Grassl, W. (2011). Hybrid forms of business: The logic of gift in the commercial world. *Journal of Business Ethics, 100*(S1), 109–123. https://doi.org/10.1007/s10551-011-1182-5

Haji, A. A., & Anifowose, M. (2016). The trend of integrated reporting practice in South Africa: Ceremonial or substantive? *Sustainability Accounting, Management and Policy Journal, 7*(2), 190–224. https://doi.org/10.1108/sampj-11-2015-0106

Hammond, N. L., Pearson, A. W., & Holt, D. T. (2016). The quagmire of legacy in family firms: Definition and implications of family and family firm legacy orientations. *Entrepreneurship Theory and Practice, 40*(6), 1209–1231. https://doi.org/10.1111/etap.12241

Haque, F. (2017). The effects of board characteristics and sustainable compensation policy on carbon performance of UK firms. *The British Accounting Review, 49*(3), 347–364. https://doi.org/10.1016/j.bar.2017.01.001

Harjoto, M. A., Laksmana, I., & Yang, Y. W. (2019). Board nationality and educational background diversity and corporate social performance. *Corporate Governance: The International Journal of Business in Society, 19*(2), 217–239. https://doi.org/10.1108/cg-04-2018-0138

Honeyman, R., & Jana, T. (2019). *The B Corp handbook: How you can use business as a force for good*. Berrett-Koehler Publishers.

Hong, B., Li, Z., & Minor, D. (2016). Corporate governance and executive compensation for corporate social responsibility. *Journal of Business Ethics, 136*(1), 199–213. https://doi.org/10.1007/s10551-015-2962-0

Hussain, N., Rigoni, U., & Orij, R. P. (2018). Corporate governance and sustainability performance: Analysis of triple bottom line performance. *Journal of Business Ethics, 149*(2), 411–432. https://doi.org/10.1007/s10551-016-3099-5

Husted, B. W., & de Sousa-Filho, J. M. (2019). Board structure and environmental, social, and governance disclosure in Latin America. *Journal of Business Research, 102*, 220–227. https://doi.org/10.1016/j.jbusres.2018.01.017

Kellermanns, F. W., Eddleston, K. A., & Zellweger, T. M. (2012). Article commentary: Extending the socioemotional wealth perspective: A look at the dark side. *Entrepreneurship Theory and Practice, 36*(6), 1175–1182. https://doi.org/10.1111/j.1540-6520.2012.00544.x

Krause, R., Withers, M. C., & Semadeni, M. (2017). Compromise on the board: Investigating the antecedents and consequences of lead independent director appointment. *Academy of Management Journal, 60*(6), 2239–2265. https://doi.org/10.5465/amj.2015.0852

Kujala, J., Sachs, S., Leinonen, H., Heikkinen, A., & Laude, D. (2022). Stakeholder engagement: Past, present, and future. *Business & Society, 61*(5), 1136–1196. https://doi.org/10.1177/00076503211066595

La Porta, R., Lopez-De-Silanes, F., Shleifer, A., & Vishny, R. W. (1998). Law and finance. *Journal of Political Economy, 106*(6), 1113–1155. https://doi.org/10.1086/250042

Le Breton-Miller, I., & Miller, D. (2006). Why do some family businesses out-compete? Governance, long-term orientations, and sustainable capability. *Entrepreneurship Theory and Practice, 30*(6), 731–746. https://doi.org/10.1111/j.1540-6520.2006.00147.x

Legifrance. (2019). *Loi n° 2019-486 du 22 mai 2019 relative à la croissance et la transformation des entreprises*. Retrieved July 11, 2024, from https://www.legifrance.gouv.fr/dossierlegislatif/JORFDOLE000037080861/

Legislation.gov.uk. (2006). *Companies Act 2006*. Retrieved July 11, 2024, from https://www.legislation.gov.uk/ukpga/2006/46

López-González, E., Martínez-Ferrero, J., & García-Meca, E. (2019). Corporate social responsibility in family firms: A contingency approach. *Journal of Cleaner Production, 211*, 1044–1064. https://doi.org/10.1016/j.jclepro.2018.11.251

Lozano, M. B., & Martínez-Ferrero, J. (2022). Do emerging and developed countries differ in terms of sustainable performance? Analysis of board, ownership and country-level factors. *Research in International Business and Finance, 62*, 101688. https://doi.org/10.1016/j.ribaf.2022.101688

Mahoney, L. S., & Thorn, L. (2006). An examination of the structure of executive compensation and corporate social responsibility: A Canadian investigation. *Journal of Business Ethics, 69*(2), 149–162. https://doi.org/10.1007/s10551-006-9073-x

Melis, A. (2000). Corporate governance in Italy. *Corporate Governance: An International Review, 8*(4), 347–355. https://doi.org/10.1111/1467-8683.00213

Michelon, G., & Parbonetti, A. (2012). The effect of corporate governance on sustainability disclosure. *Journal of Management & Governance, 16*(3), 477–509. https://doi.org/10.1007/s10997-010-9160-3

Miller, D., & Le Breton-Miller, I. (2014). Deconstructing socioemotional wealth. *Entrepreneurship Theory and Practice, 38*(4), 713–720. https://doi.org/10.1111/etap.12111

Morck, R., & Yeung, B. (2004). Family control and the rent-seeking society. *Entrepreneurship Theory and Practice, 28*(4), 391–409. https://doi.org/10.1111/j.1540-6520.2004.00053.x

Naciti, V., Cesaroni, F., & Pulejo, L. (2022). Corporate governance and sustainability: A review of the existing literature. *Journal of Management and Governance, 26*(1), 55–74. https://doi.org/10.1007/s10997-020-09554-6

NATIVA. (2024). *The AIDAF legacy book: How to be a good ancestors*. Retrieved July 7, 2024, from https://nativalab.com/stories/il-legacy-book-di-aidaf-come-essere-buoni-antenati/

References

Noland, J., & Phillips, R. (2010). Stakeholder engagement, discourse ethics and strategic management. *International Journal of Management Reviews, 12*(1), 39–49. https://doi.org/10.1111/j.1468-2370.2009.00279.x

OECD. (2023). *OECD corporate governance Factbook 2023*. OECD Publishing.

OECD. (2024). *Global corporate sustainability report 2024*. OECD Publishing.

Orazalin, N. (2020). Do board sustainability committees contribute to corporate environmental and social performance? The mediating role of corporate social responsibility strategy. *Business Strategy and the Environment, 29*(1), 140–153. https://doi.org/10.1002/bse.2354

Pozzoli, M., Pagani, A., & Paolone, F. (2022). The impact of audit committee characteristics on ESG performance in the European Union member states: Empirical evidence before and during the COVID-19 pandemic. *Journal of Cleaner Production, 371*, 133411. https://doi.org/10.1016/j.jclepro.2022.133411

Rees, W., & Rodionova, T. (2015). The influence of family ownership on corporate social responsibility: An international analysis of publicly listed companies. *Corporate Governance: An International Review, 23*(3), 184–202. https://doi.org/10.1111/corg.12086

Ricart, J. E., Rodríguez, M. A., & Sánchez, P. (2005). Sustainability in the boardroom: An empirical examination of Dow Jones sustainability world index leaders. *Corporate Governance: The International Journal of Business in Society, 5*(3), 24–41. https://doi.org/10.1108/14720700510604670

Semenova, N., & Hassel, L. G. (2019). Private engagement by Nordic institutional investors on environmental, social, and governance risks in global companies. *Corporate Governance: An International Review, 27*(2), 144–161. https://doi.org/10.1111/corg.12267

Sharma, P., & Sharma, S. (2011). Drivers of proactive environmental strategy in family firms. *Business Ethics Quarterly, 21*(2), 309–334. https://doi.org/10.5840/beq201121218

Shaukat, A., Qiu, Y., & Trojanowski, G. (2016). Board attributes, corporate social responsibility strategy, and corporate environmental and social performance. *Journal of Business Ethics, 135*(3), 569–585. https://doi.org/10.1007/s10551-014-2460-9

Sustainability Accounting Standards Board. (2011). *SASB Standards*. Retrieved July 8, 2024 from https://sasb.ifrs.org

Suttipun, M. (2021). The influence of board composition on environmental, social and governance (ESG) disclosure of Thai listed companies. *International Journal of Disclosure and Governance, 18*(4), 391–402. https://doi.org/10.1057/s41310-021-00120-6

Velte, P. (2024). Archival research on sustainability-related executive compensation. A literature review of the status quo and future improvements. *Corporate Social Responsibility and Environmental Management, 31*(4), 3119–3147. https://doi.org/10.1002/csr.2741

Venturelli, A., Principale, S., Ligorio, L., & Cosma, S. (2021). Walking the talk in family firms. An empirical investigation of CSR communication and practices. *Corporate Social Responsibility and Environmental Management, 28*(1), 497–510. https://doi.org/10.1002/csr.2064

Walker, D. H. T., Bourne, L. M., & Shelley, A. (2008). Influence, stakeholder mapping and visualization. *Construction Management and Economics, 26*(6), 645–658. https://doi.org/10.1080/01446190701882390

Webb, E. (2004). An examination of socially responsible firms' board structure. *Journal of Management and Governance, 8*(3), 255–277. https://doi.org/10.1007/s10997-004-1107-0

Zattoni, A. (2019). The evolution of corporate governance in Italy: Formal convergence or path-dependence? *Corporate Governance and Research & Development Studies, 1*, 13–35. https://doi.org/10.3280/cgrds1-2019oa8799

Zellweger, T. M., Nason, R. S., Nordqvist, M., & Brush, C. G. (2013). Why do family firms strive for nonfinancial goals? An organizational identity perspective. *Entrepreneurship Theory and Practice, 37*(2), 229–248. https://doi.org/10.1111/j.1540-6520.2011.00466.x

Zhang, L. (2012). Board demographic diversity, independence, and corporate social performance. *Corporate Governance: The International Journal of Business in Society, 12*(5), 686–700. https://doi.org/10.1108/14720701211275604

Chapter 4
Management and Governance of Family Business

4.1 Overview of the Family Business Phenomenon

Family firms constitute the most common organizational structure in the world (La Porta et al., 1999) and significantly impact the world economy (Astrachan & Shanker, 2003; Morck & Yeung, 2004).

As noted by the 2023 EY and University of St. Gallen Family Business Index,[1] the largest 500 family businesses in the world outpace the global economy, growing at nearly twice the rate of advanced economies and approximately one and a half times the rate of emerging markets and developing economies. Together, these businesses produce $8.02 trillion in revenue and employ 24.5 million people globally,

---

[1] The Family Business Index is a ranking of the 500 largest family businesses in the world by revenue. It is issued every two years. The criteria for entering the Index are as follows:

- The business should be in the second generation or more. If the business is still in the first generation, at least two family members should be in the board of directors, supervisory board, or the executive leadership team.
- The family should have substantial ownership and thus decision-making authority in the business. Private companies are family firms in case the family controls (directly or through a family foundation or trust) at least 50% of the voting rights. Public listed companies are family firms in case the family holds (directly or through a family foundation or trust) at least 32% of the voting rights.
- To allow wider representation of families, the Index features one business per family. If two or more companies controlled by the same family qualify for the Index, the Index will feature either the parent/holding company or the business with the highest revenues.
- Revenues of featured businesses should originate from published accounts that are no more than 24 months old.
- Data sources used to generate the Index include public domain, filed and published financial statements or annual reports, and commercially accessible databases including Bloomberg, S&P CapIQ, Orbis, BoardEx, and D&B Hoovers.

© The Author(s), under exclusive license to Springer Nature Switzerland AG 2025
F. R. Arduino, *Governance and Sustainability of Family Business*, International Series in Advanced Management Studies,
https://doi.org/10.1007/978-3-032-01944-8_4

representing a 10% and 1.4% increase, respectively, compared with the 2021 Index (EY and University of St. Gallen Family Business Index, 2023).

The 2023 Index includes a slightly greater number of public companies than private companies, with 260 publicly listed family businesses on the list, an increase from 243 in 2017.

The 500 companies featured in the Index are located in 47 different jurisdictions. Nearly half of these businesses are located in Europe. North America accounts for 30% of family businesses, while the Asia–Pacific region represents 16%. Geographically, as illustrated in Table 4.1, the U.S. dominates the family business landscape, with 119 companies in the Top 500. Conversely, seven of the ten largest family businesses in the world are headquartered in the U.S. These firms collectively generate $2.7 trillion in revenue and employ 6.3 million people.

Europe has a long tradition of family businesses, which are crucial to the EU economy. These businesses vary widely, from sole proprietorships to large multinational corporations, from publicly traded to privately held, and they account for more than 60% of all enterprises in Europe (EY and University of St. Gallen Family Business Index, 2023). Germany has the most family businesses in Europe, representing the second-largest contributor to the 2023 EY and University of St. Gallen Family Business Index. Specifically, Germany has 78 companies on the list, which together generate $1.13 trillion in revenue and employ 3.35 million people.

In terms of industry, as shown in Fig. 4.1, the 2023 EY and University of St. Gallen Family Business Index reveals that the consumer sector represents the largest sector at 37.4%, followed by the advanced manufacturing and mobility (AM&M) sector, which, compared with the past, has expanded its presence on the Index to 28.6%.

Over three-quarters (76%) of the companies on the 2023 Index have been established for more than 50 years, and nearly one-third (31%) have existed for more than a century. The longevity of family firms highlights their ability to sustain success and effectively manage intergenerational transitions across time.

Table 4.1 The top ten jurisdictions across the globe by revenue in the 2023 EY and University of St. Gallen Family Business Index

Country	Revenue (US$ billion)	Number of businesses
United States	2721.9	119
Germany	1125.8	77
France	503.2	31
India	364.6	15
South Korea	340.5	16
China	245.6	11
Canada	243.7	16
Hong Kong	240.1	18
Switzerland	235.3	15
Netherlands	194.2	13

Source: EY and University of St. Gallen Family Business Index. Adapted from: https://familybusinessindex.com/

4.1 Overview of the Family Business Phenomenon

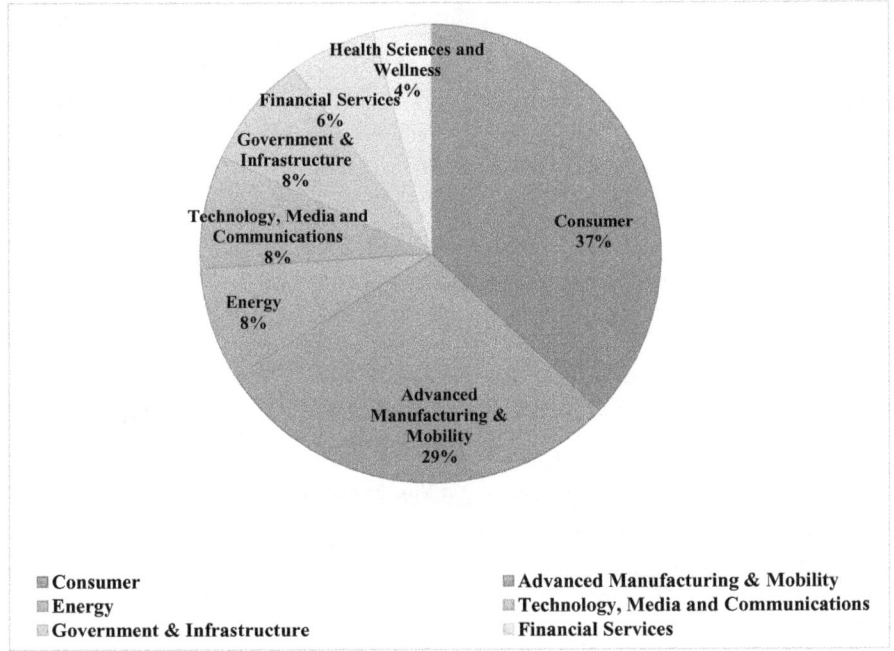

Fig. 4.1 Revenue Breakdown of Family Businesses by Industry in the 2023 EY and University of St. Gallen Family Business Index. Source: Author's elaboration on the EY and University of St. Gallen Family Business Index (https://familybusinessindex.com/)

Examining the governance of family businesses, the 2023 Index confirms that family members are actively involved in leadership and management roles. Approximately one-quarter (23%) of all board seats across the Index are occupied by family members. Additionally, nearly half of the companies in the Index (45%, i.e., 223 companies) have a family member serving as CEO. Interestingly, as shown in Fig. 4.2, older companies are less likely to have family members in leadership positions, a trend that is also observed among the youngest companies in the Index. Moreover, 19% of family businesses include a member of the family's next generation (aged 40 years or younger) on their board. The average age of a family board member in the 2023 Index has increased to 62 years, up from 60 in 2021.

Although younger board members can bring fresh expertise and knowledge and their role is crucial in addressing the challenges posed by social, environmental, economic, and technological transformations, family businesses need to do more to ensure diversity and inclusion. The 2023 Index shows that women occupy 23% of board seats across the largest 500 family businesses worldwide, and Europe is leading, as women hold 25% of board seats. Additionally, in terms of leadership, only 29 companies (5.8% of the total) have a female CEO, which represents a slight increase from 2021 when 27 companies were led by women.

Considering the Italian presence in the Index, the first two Italian family companies rank 38 and 120, respectively, in the list. As illustrated by Table 4.2, the first ten

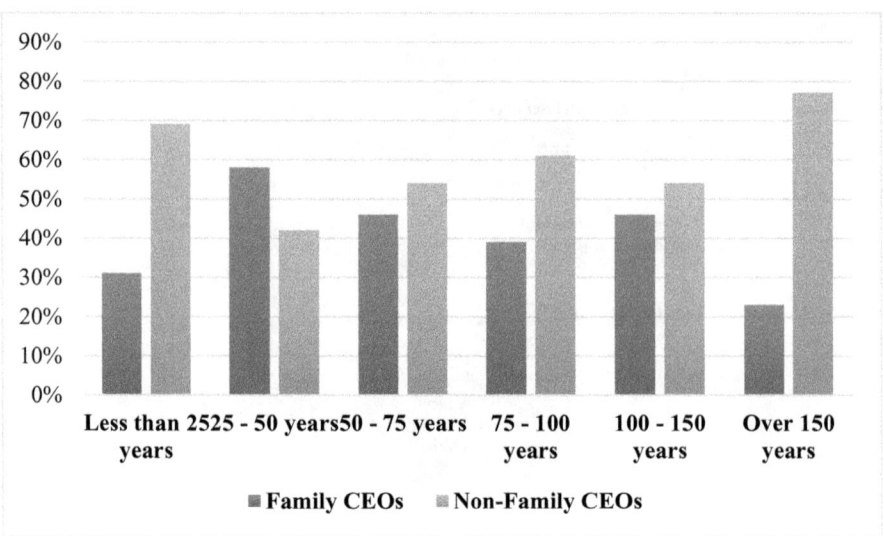

Fig. 4.2 Presence of Family CEOs by Age of Family Businesses in the 2023 EY and University of St. Gallen Family Business Index. Source: Author's elaboration on the EY and University of St. Gallen Family Business Index (https://familybusinessindex.com/)

Table 4.2 The first ten Italian family businesses in the 2023 EY and University of St. Gallen Family Business Index

Company	Type	Founding year	Family	Shareholding (%)	Industry	Revenue (US$ billion)
Exor SpA	Public	1927	Agnelli	53	Financial services	39.8
Ferrero International SA	Private	1946	Ferrero	100	Consumer	14.9
Edizione Srl	Private	1955	Benetton	100	Consumer	11.1
Saras Group	Public	1962	Moratti	40.2	Energy	10.1
Esselunga SpA	Private	1957	Caprotti	100	Consumer	10.1
Marcegaglia SpA	Private	1959	Marcegaglia	100	Energy	9.5
Webuild SpA	Public	1906	Salini	45.2	Government & infrastructure	7.1
Italiana Petroli SpA	Private	1933	Peretti	100	Energy	5.3
Chimet SpA	Private	1976	Squarcialupi	72	Energy	5.2
Gruppo Barilla	Private	1877	Barilla	100	Consumer	4.7

Source: The EY and University of St. Gallen Family Business Index. Adapted from: https://familybusinessindex.com/

4.1 Overview of the Family Business Phenomenon

Italian family businesses in the Index are either public or private, they belong to different industries, and they are all characterized by family shareholdings greater than 40%.

According to the 2023 permanent census by the Italian National Institute of Statistics (ISTAT, 2023), in Italy, there is a strong presence of companies controlled by an individual or a family (more than 820,000 units), representing 80.9% of all companies with at least three employees (compared with 75.2% in 2018). This phenomenon is particularly widespread among microenterprises (83.3% of cases) and less common among small (74.5%), medium (58.8%), and large enterprises (41.6%).

In terms of industry, family businesses are more prevalent in the manufacturing sector (81.2%), particularly in traditional sectors such as textiles, clothing, footwear, food, and wood; in the construction sector (82.4%); and in services, especially in trade (84.4%) and accommodation and food services (87.3%).

With respect to governance, the management of these family businesses is typically handled by the entrepreneur or a member of the owning family. However, internal or external managers are employed more frequently in medium-sized (10.4% of the units considered) and large enterprises (21.3%). Between 2016 and 2022, just under one in ten companies reported having undergone at least one generational transition. Among medium and large enterprises, these percentages increase to 17.8% and 18.9%, respectively.

According to the 15th edition of the AUB Observatory (AUB Observatory XV Edition, 2024), 59% of Italian family businesses have at least one nonfamily member on the board. Additionally, women represent more than one-third of the members (i.e., 33%) on the boards of 38.1% of companies. Finally, 93.4% of the firms have boards with fewer than two members older than 75, whereas 26.2% of the firms have at least one board member younger than 40. This evidence suggests that Italian family businesses struggle to transition toward more evolved governance models.

With respect to leadership structure, 27.4% of Italian family businesses are governed by a sole director. Other common leadership structures include an executive chairperson, a single CEO, or collegial leadership. Family leadership models remain predominant, although they have become less prevalent in both large and small companies over the past decade. The transition from family leaders to nonfamily leaders has been more common in smaller companies.

4.2 Definition and Distinctive Features

The term "family business" lacks a consistent and clear definition, largely because it encompasses a wide variety of characteristics and organizational structures. A range of definitions can be drawn from various sources, including scientific literature, empirical research, and supranational organizations.

Over time, the interest in family firms has grown significantly. Various organizations that are specifically dedicated to the study and advanced analysis of this phenomenon, including the International Family Enterprise Research Academy (IFERA), the Family Firm Institute, the Family Business Network International, and several others, have emerged. In academia, numerous scientific journals have significantly expanded their coverage of family business studies, and many scholars have examined family businesses (Machek, 2016). These investigations have produced a variety of definitions derived from the diverse perspectives and insights that have emerged from their research.

Davis (2001) categorized the different definitions of family business into two primary groups: process definitions and structural definitions. The process definition emphasizes the degree of involvement that a family has in business operations. In contrast, the structural definition focuses primarily on the ownership or management structures within a family business.

Morck et al. (1988) defined a company as a family firm if one of the top two executives is a member of the founding family. According to Astrachan et al. (2002), a family firm is characterized by significant family ownership, the presence of family members in governing bodies and key managerial roles, active family involvement in management, and the intent to keep the business within the family across generations, highlighting a long-term orientation. According to Anderson and Reeb (2003), a company qualifies as a family firm if the founding family has fractional equity ownership and/or family members are present on the board of directors. Chrisman et al. (2005) classified businesses as family businesses if they are governed and/or managed with the intent of shaping and pursuing the business vision held by a dominant coalition controlled by members of the same family or a small number of families, aiming for sustainability across multiple generations of the family or families. According to Fahlenbrach (2009), a company can be defined as a family firm if the CEO is the founder or cofounder, whereas according to Lane et al. (2006), a family firm is characterized by ownership and control by a single family, the active involvement of family members in the governance and management of the business, and a clear intention to maintain family control and influence across multiple generations. Bennedsen et al. (2007) defined a family firm as one in which the incoming CEO is related by blood or marriage to the outgoing CEO. As stated by Miller et al. (2007), a family business is characterized by the involvement of multiple family members as major owners or managers, either concurrently or across different periods.

Many definitions vary on the basis of the proportion of shareholdings or voting rights owned by the family. According to Claessens et al. (2000), family groups are characterized by their control of more than 5% of their company's voting rights. Gómez-Mejía et al. (2003) defined a family-controlled firm on the basis of two criteria: having two or more directors with a family relationship and having family members who own or control at least 5% of the voting stock. According to Villalonga and Amit (2006), a company can be classified as a family firm if the founder or a family member serves as an officer or director or owns more than 5% of the company's equity. Maury (2006) considered a company a family firm when the largest controlling shareholder who holds at least 10% of the voting rights is a family, an individual, or an unlisted firm. As noted by Barontini and Caprio (2006), a family

4.2 Definition and Distinctive Features

firm is defined as one where the largest shareholder possesses at least 10% ownership rights and either the family or the largest shareholder controls more than 51% of the direct voting rights or holds more than twice the direct voting rights of the second largest shareholder. According to La Porta et al. (1999), a company qualifies as a family firm if a person is the controlling shareholder (ultimate owner) whose direct and indirect voting rights exceed 20%. Barth et al. (2005) classified a company as a family firm if one person or a single family owns at least 33% of the firm's shares, whereas Ang et al. (2000) defined a family firm as one where a single family controls more than 50% of the firm's shares. According to Holderness and Sheehan (1988), a company can be categorized as a family firm if a single majority shareholder or entity holds at least 50.1% of the company's stock, which may also include trusts and foundations.

In Europe, a European Commission study revealed over 90 different definitions of family businesses used by scholars, institutions, and research centers across various member states. In 2009, the European Commission developed a working definition of family business, which was established to provide a common understanding and to facilitate the comparison of family businesses across EU member states. According to the European Commission, a family business is characterized by the following criteria (European Commission, 2009, p. 10):

(i) "The majority of decision-making rights are in the possession of the natural person(s) who established the firm, or in the possession of the natural person(s) who has/have acquired the share capital of the firm, or in the possession of their spouses, parents, child, or children's direct heirs.
(ii) The majority of decision-making rights are indirect or direct.
(iii) At least one representative of the family or kin is formally involved in the governance of the firm.
(iv) Listed companies meet the definition of family enterprise if the person who established or acquired the firm (share capital) or their families or descendants possess 25 per cent of the decision-making rights mandated by their share capital."

The European Commission has highlighted several key challenges faced by businesses that are categorized as family firms. These challenges include the following:

– The need to plan and manage business transfers;
– Difficulties in accessing finance and dealing with taxation issues, despite the crucial role that small- to medium-sized firms play in the economic growth of their respective countries;
– Challenges in family governance, particularly in achieving a balance between family dynamics, ownership interests, and economic objectives. Personal family goals sometimes conflict with the financial and economic goals that businesses must pursue;
– Issues related to attracting and retaining a skilled workforce;
– The need for entrepreneurship education and specialized management training that are tailored to the unique dynamics of family-owned businesses.

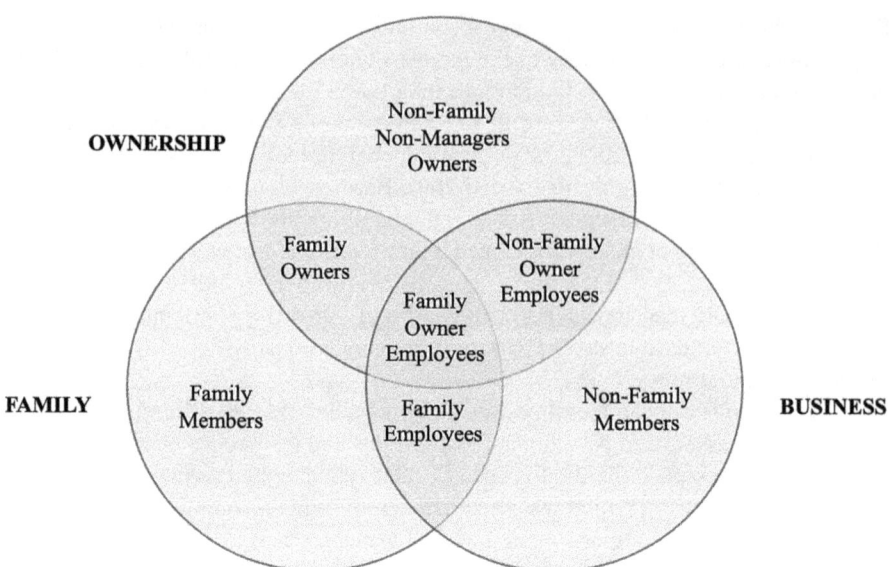

Fig. 4.3 The Three-Circle Model of the Family Business System. Source: Author's elaboration on Tagiuri and Davis (1996)

Family firms exhibit intricate overlap among family, business, and ownership. This complexity can potentially lead to conflicts arising from the diverse perspectives and interests of the various stakeholders involved. Tagiuri and Davis (1996) developed the three-circle model of the family business system, a conceptual framework that illustrates the complex relationships and dynamics within family businesses. As shown in Fig. 4.3, this model consists of three overlapping circles representing family, business, and ownership. Each circle encompasses different roles and interests, and their intersections highlight the various positions that individuals can occupy within the system. The Family circle includes all family members, regardless of whether they are involved in the business or own shares; the Business circle represents the employees and managers working within the business; and the Ownership circle includes all of the individuals who have an ownership stake in the business. This model is helpful for identifying potential sources of conflict, such as role ambiguity and competing interests, and highlights the importance of proper governance structures and succession planning to ensure long-term sustainability and harmony within the family business system.

Astrachan et al. (2002) established an index of family influence (F-PEC) by assessing three critical dimensions that can be used to measure the extent of family involvement: power, experience, and culture (Fig. 4.4). Power indicates the extent of family members' engagement in ownership and management. Experience measures the involvement of multiple generations in both ownership and management, providing an indicator of the continuity of resources across time and the commitment to business legacy. Culture is used to assess the alignment between corporate

4.2 Definition and Distinctive Features

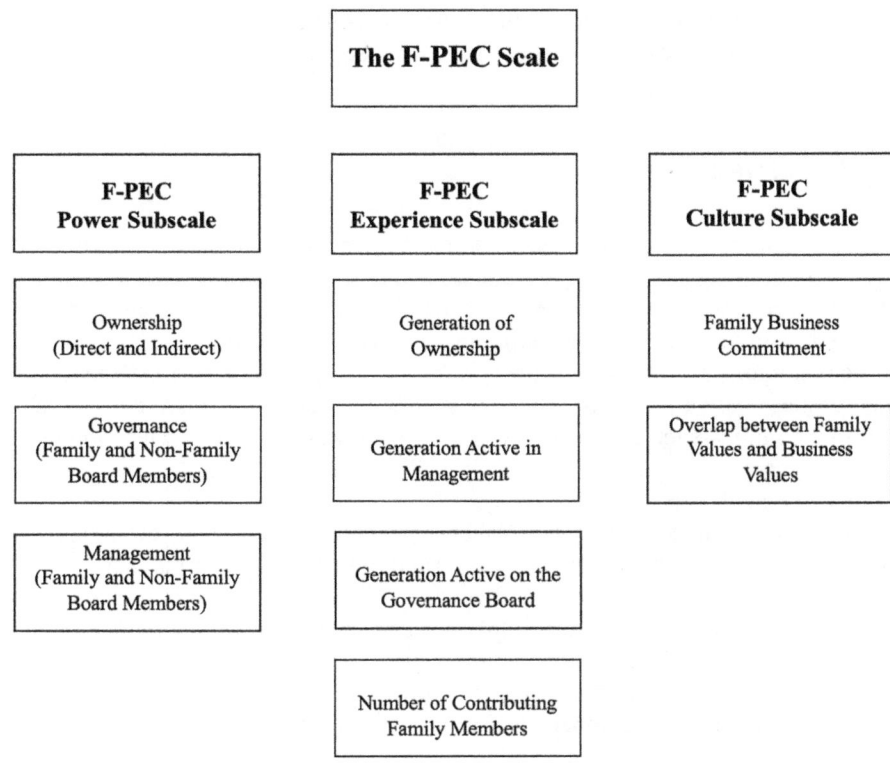

Fig. 4.4 The F-PEC Scale Index. Source: Adapted from Astrachan et al. (2002)

and family values, as well as the dedication of family members to supporting management in achieving business goals and promoting harmonious relationships among the family, the organization, and the broader environment. Therefore, the F-PEC Scale Index provides a comprehensive measure of family influence in a family business, which includes both tangible aspects such as ownership (direct and indirect) and board representation (i.e., presence of family members on the board) and intangible aspects such as cultural alignment and generational involvement in management or governance.

4.3 Main Theoretical Frameworks

Since a universally accepted model establishing the criteria that define a family business is lacking, various conceptual frameworks need to be examined. Each framework highlights distinct aspects of the phenomenon, delineating its characteristics and illustrating fundamental differences from nonfamily firms. These models highlight not only the unique dynamics of family involvement in business governance and ownership but also the complexities involved in defining and examining family businesses in a comprehensive manner.

4.3.1 Agency Theory

In agency theory, the principal–agent problem arises from the divergence of interests between the owner (principal) and the managers (agents) in charge of running the business. This conflict typically results in agency costs, which are expenses incurred in monitoring and controlling the actions of agents to ensure that they act in the best interests of the owners (Jensen & Meckling, 1976). In family businesses, this conflict is often mitigated when ownership and management are consolidated within the family, such as when a family member serves as CEO. This alignment reduces agency costs because the interests and objectives of ownership and management are aligned.

According to agency theory, optimal business outcomes are achieved when ownership and management are aligned. Compared with nonfamily firms, family businesses are ideally positioned to minimize the principal-agent problem since family members, who are driven by emotional ties and interest in the family legacy, are typically more dedicated to reducing these costs. This dedication increases when family members in management positions are selected on the basis of merit rather than nepotism, ensuring competence and reducing conflicts of interest (Villalonga et al., 2015).

Conflicts can arise from asymmetric altruism in family firms (Schulze et al., 2001). This can determine an agency threat when owners or managers use firm resources excessively to benefit their family. Generosity can encourage free-riding and nepotism, contributing to entrenchment tendencies (Jaskiewicz et al., 2013). Schulze et al. (2001) hypothesized that higher board entrenchment correlates with lower performance in privately held, family-managed firms, whereas Morck and Yeung (2003) asserted that entrenchment increases agency conflicts more in family firms than in nonfamily firms. However, not all scholars view altruism as a source of agency conflict in family firms. For example, Chrisman et al. (2004) argued that family firms may pursue both economic and noneconomic objectives and that agency costs occur only when managers prioritize their own interests over those of the owners. Therefore, according to their perspective, if a family business prioritizes hiring family members regardless of their qualifications, this should not be considered an agency problem. However, altruism in family firms can play a significant role in the reduction of agency costs (Schulze et al., 2003). Unlike nonfamily businesses, in which managers may prioritize their personal interests or short-term gains, family members in managerial roles often exhibit altruistic behaviors toward the family and the business. Therefore, altruism contributes to lower agency costs in family firms by promoting behaviors and decision-making processes that are aligned with the long-term interests of the business and its stakeholders.

The second agency problem arises when controlling shareholders (typically the family) and minority shareholders have a conflict of interest. In family businesses, where large shareholders often exert significant influence, the risk of expropriation or neglecting minority shareholders' interests is greater. Conversely, in professionally managed firms with dispersed ownership, this conflict is less pronounced (Villalonga et al., 2015). Moreover, family businesses tend to use debt financing to retain control rather than dilute ownership, further complicating agency dynamics with creditors.

Another agency issue in family businesses involves conflicts between the overarching family interests and the goals of individual family members, who are shareholders. The family's collective aspirations, such as preserving the family's reputation, maintaining unity, and ensuring the business' continuity across generations, often diverge from the personal objectives of individual shareholders. This dual focus can lead to internal conflicts and challenges in balancing familial and business priorities.

Additionally, family businesses face unique challenges related to executive entrenchment, which involves managers entrenching themselves to secure their positions by manipulating information or adopting strategies that consolidate their power (Gómez-Mejía et al., 2001). This behavior can undermine corporate governance, impacting a firm's long-term sustainability.

Conflicts can also arise among family members occupying different roles within the business, such as owners versus managers or board members, which may lead to disagreements over strategic decisions and operational policies (Anderson & Reeb, 2003). These conflicts are exacerbated when family members lack the necessary skills or exhibit free-riding behaviors, which leads to efficiency being reduced and effective collaboration being hindered.

Agency theory has been frequently employed in prior research to examine the governance structures and sustainability practices in the context of family firms. By highlighting the potential conflicts of interest between family owners or managers and various stakeholders, agency theory provides a useful lens through which to understand how governance mechanisms can either support or hinder sustainability-oriented strategies within family firms (El Ghoul et al., 2016).

According to agency theory, family firms may prioritize the interests of the controlling family over those of other stakeholders. In this context, the board of directors plays a crucial governance role by mitigating agency problems specific to family firms, such as asymmetric altruism and managerial entrenchment (Corbetta & Salvato, 2004a; Eddleston & Kidwell, 2012). The board's oversight is essential to ensuring that managerial decisions align with the broader interests of the principal, rather than being solely influenced by family-centered goals (Chrisman et al., 2007). Acting on behalf of stakeholders, the board performs a critical monitoring function (Aguilera et al., 2006), helping to balance financial performance with the pursuit of social responsibility objectives.

The emphasis on family-centered goals can impede a firm's commitment to corporate social responsibility, as decisions around sustainability investments may generate conflicts of interest between managers, shareholders, and other stakeholders.

Rees and Rodionova (2015) argue that family members are often resistant to extensive investment in ESG practices, particularly when such initiatives are perceived as offering limited personal benefit. This resistance may stem from a preference for strategies that maximize personal returns or from a belief that ESG investments are potentially detrimental to firm value.

4.3.2 Resource-based View Theory

Resource-based view (RBV) theory posits that firms are made up of a collection of resources, with some being rare, valuable, nonimitable, and nonsubstitutable. These unique resources give firms a competitive advantage, which translates into superior performance (Barney, 1991). In the context of family firms, the family itself can be considered such a resource (Rau, 2014).

According to Habbershon and Williams (1999), family firm resources, termed *familiness*, arise from the interaction among the business, the family, and the firm's members. This concept provides a framework for connecting performance outcomes to firm characteristics, emphasizing traits that make resources in family firms valuable, rare, inimitable, and organizationally embedded (VRIO). Key traits contributing to VRIO resources in family firms include path dependence and causal ambiguity. Path dependence is related to historical conditions that are unique to family firms and create nonimitable resources. These conditions include a firm's culture, geographic location, historical assets, reputation, informal decision-making processes, and mentoring between family generations. Causal ambiguity refers to an unclear understanding of the relationship between firm resources and competitive advantage. Families often do not recognize the resources that create their competitive advantage, which becomes evident only during periods of change or difficulty.

According to Sirmon and Hitt (2003), family firms have the following unique resources and attributes that give them potential advantages over nonfamily firms:

- Human capital: this capital reflects the intricacy of the knowledge, skills, and abilities of a person or an organization. In a family firm, human capital is endowed with a sort of added value derived from the fact that each family member, through participation in both business and family life, has and brings original and inimitable elements.
- Social capital: social capital expresses the complexity of current and potential resources derived from the set of relationships created between a plurality of individuals and organizations. In a family firm, the network of relationships that is created between family members and stakeholders favors the creation of stable and productive links between the business and stakeholders themselves.
- Patient financial capital: family firms tend to keep resources in the company for an extended period, therefore, reducing the risk of funds being subject to repayment. This element favors the investment and growth of the company.

4.3 Main Theoretical Frameworks

- Survivability capital: this capital represents the set of resources that family members decide to bring personally (working in the company without remuneration and granting personal loans). These resources are typically exploited with greater intensity at the start of a business and in times of business difficulty.
- Governance structure: due to their strong family and fiduciary ties, family firms may need fewer expensive measures to mitigate agency costs, such as extensive management monitoring and performance-related compensation tools.

Adopting a social capital perspective, Pearson et al. (2008) proposed a three-dimensional model to identify the behavioral and social resources constituting *familiness*:

- The Structural dimension involves social interactions and the strength of family members' network ties, which provide a firm with unique social capital;
- The Cognitive dimension refers to the shared vision, language, and values within a company, which are rooted in the family's history and culture and favor a common identity;
- The Relational dimension includes trust, norms, obligations, and identification within a family, which create unique and lasting bonds that influence behavior.

Since family firms involve the integration of family and business dimensions, their structure is unique, differentiating them from other types of enterprises. This uniqueness, *familiness*, is a key factor in their competitive advantage and is driven by the positive involvement and influence of family members in business activities. The resources and capabilities that support *familiness* contribute to explaining the competitive edge of family businesses and their approach to seizing opportunities.

The resource-based view theory offers a valuable lens through which to understand how the unique characteristics and internal resources of family firms influence their sustainability performance. For instance, Huang et al. (2016) combine the RBV with behavioral theory to investigate how family influence and firm-specific factors affect the adoption of green product innovations. Similarly, Veiga (2025) applies the RBV to analyze how family firms can leverage their unique resource base to develop environmentally sustainable practices and products.

The distinctive traits of family firms enhance their ability to implement effective sustainability strategies. Family firms can strategically leverage their reputational capital (Deephouse & Jaskiewicz, 2013; Yang & Basile, 2022) to achieve superior sustainability outcomes, and in turn, CSR activities can serve to further build and reinforce that reputation among external stakeholders (Stock et al., 2024). In addition, the social capital embedded in the family (Arregle et al., 2007; Sanchez-Ruiz et al., 2019) plays a crucial role in driving engagement with key nonfamily stakeholders, including customers, employees, and the broader community (Cabrera-Suárez et al., 2015).

4.3.3 Socioemotional Wealth Theory

SEW theory, which was developed by Gómez-Mejía et al. (2007), posits that family businesses prioritize the nonfinancial aspects of their wealth that are tied to the family's emotional and social needs. This focus on SEW influences their decision-making processes and strategic choices. To comprehensively understand the nonfinancial factors shaping decisions and behaviors in family firms, Berrone et al. (2012) introduced the FIBER framework. This framework delineates five dimensions of SEW that are specific to family businesses:

- Family control and influence (F): this dimension emphasizes the importance of family members retaining control and significant influence over the strategic decisions and operations of their business. Family control ensures that the firm's direction aligns with the family's values, goals, and vision. This influence often extends to key management positions, board membership, and decision-making processes.
- Identification of family members with the firm (I): family members often view their business as an extension of themselves, which leads to a strong sense of identification with the firm. This identification promotes deep emotional attachment and a personal commitment to their company's success. It also helps create a strong corporate culture that reflects the family's values and traditions.
- Binding social ties (B): social relationships and networks within and outside the family play crucial roles in family businesses. Binding social ties refer to the close, personal relationships that family members have with each other and with other stakeholders, such as employees, customers, suppliers, and the community. These ties create a sense of loyalty, trust, and mutual support, which can be a significant source of competitive advantage.
- Emotional attachment (E): this reflects family members' affective commitment to their business. This dimension highlights the emotional investment that family members have in the firm, which often goes beyond financial considerations. Emotional attachment can drive family members to prioritize the long-term sustainability of the business, make personal sacrifices for its benefit, and ensure its continuity across generations;
- Renewal of family bonds to the firm through dynastic succession (R): the desire to pass on the business to future generations is a key aspect of SEW. This dimension is centered on the long-term vision of maintaining the family legacy and ensuring that the firm remains under family control through dynastic succession. Planning for succession within the family is crucial for preserving SEW and maintaining the continuity of family influence and values in the business.

According to Le Breton-Miller and Miller (2013), the stages of evolution in family firms and the varying degrees of family involvement can shape owners and managers' SEW priorities, and these can, in turn, determine the board composition necessary to increase a firm's survival. Le Breton-Miller and Miller identified three stages of evolution in family firms:

4.3 Main Theoretical Frameworks

- Founder firms: these are young and small firms that are typically owned and run by their founder and have strong family commitment but limited resources. Their priority is passing a healthy firm to future generations. In terms of board composition, boards should include resource suppliers (also other small entrepreneurs) and committed family members who provide their contributions at low costs;
- Postfounder family firms: these firms are established and medium-sized, with continued family ownership and management. In this stage, emotional attachment to the firm and the desire to preserve ownership remain strong. These firms prioritize family rewards, often experiencing conflict, particularly when varying levels of involvement among family members occur. For board composition, it is beneficial to have equitable family and gender representation to ensure fairness. Senior family members can help reduce tensions, sustain the founder's vision, and resolve conflicts (Ward, 2004). Additionally, directors from the local community can provide valuable support;
- Cousin consortia: these are older, larger, and more complex firms that are often characterized by dispersed family ownership and nonfamily management. Governance becomes increasingly complicated, increasing the potential for conflicting agendas among family members. Priorities shift toward utilizing the business primarily for financial support (Miller et al., 2013), with reduced family identification and emotional attachment. With respect to board composition, family representation, which enhances harmony and maintains interest in the business, needs to be balanced with qualified, experienced, and objective members who can manage conflicts and provide essential expertise.

Succession planning in family firms is heavily influenced by SEW (Minichilli et al., 2014). Family businesses are more likely to choose successors from within the family, even if external candidates might be more qualified. The emotional ties and desire to maintain the family legacy often outweigh purely economic considerations in these decisions. While this focus on internal succession can create challenges, it can also promote a sense of continuity and stability.

SEW theory also has significant implications for decision-making in family businesses. Some scholars have argued that a focus on nonfinancial goals can lead to suboptimal business decisions that compromise profitability. An emphasis on SEW can lead to conservative financial strategies and risk aversion. Family firms might be reluctant to diversify or seek external capital if it threatens their control and influence (Gómez-Mejía et al., 2010, 2011). This can lead to opportunities for growth and innovation being missed, but it also ensures the preservation of the family's emotional and social investments in the business. According to Sciascia et al. (2014), family management tends to positively impact profitability in later generational stages, as the reduced emphasis on preserving SEW encourages family managers to prioritize increasing financial wealth.

SEW theory has emerged as a central framework for understanding the unique behaviors and strategic decisions of family firms in relation to sustainability, offering insight into why these firms may adopt socially and environmentally responsible practices. Scholars have applied SEW to explore how the desire to safeguard the family's reputation, ensure transgenerational continuity, and maintain emotional ties to the firm often leads family firms to engage in long-term and sustainability-oriented strategies. One of the main arguments in studies influenced by SEW theory is that family firms may engage in socially responsible behavior to protect and enhance their image and reputation, therefore avoiding actions that could be perceived as irresponsible or damaging by stakeholders (e.g., Dyer & Whetten, 2006; Labelle et al., 2018; Dick et al., 2021).

Le Breton-Miller and Miller (2016) highlight that the development of socially responsible owners and managers in family firms is influenced by values taught within the family, strong parenting practices, and positive educational experiences. According to Marques et al. (2014), firms exhibiting higher degrees of family involvement tend to demonstrate a greater propensity to engage in sustainability-oriented initiatives. Family firms have been shown to engage in more environmentally responsible behavior, such as lower levels of pollution (Berrone et al., 2010; Berrone et al., 2023; Gómez-Mejía et al., 2025), and to maintain stronger relationships with stakeholders (Cennamo et al., 2012). Additionally, their deep embeddedness within immediate social networks further reinforces a strong sense of responsibility and commitment to socially responsible actions (Lumpkin & Bacq, 2022).

However, according to some scholars (Kellermanns et al., 2012; Zientara, 2017), the inherently ambiguous nature of SEW can lead to detrimental outcomes for stakeholders in family firms. In such cases, corporate social responsibility may be treated selectively and instrumentally as a means of securing self-interested gains under the guise of preserving SEW, rather than being embraced as a core strategic priority driven by genuine concern for the broader social good.

At the same time, SEW theory also helps explain instances of risk aversion or resistance to change when sustainability efforts are perceived to threaten socioemotional priorities. For instance, the pursuit of sustainable innovation and investments in R&D may be constrained by the desire to preserve SEW (Chrisman & Patel, 2012).

4.3.4 Stewardship Theory

Donaldson and Davis (1991) introduced stewardship theory as a counterpoint to agency theory, which was criticized for overlooking the positive social relationships that can exist between owners and managers, especially in family businesses. Davis et al. (1997) further developed the stewardship concept to address the limitations of agency theory; unlike agency theory, which assumes a potential conflict between the interests of managers and owners, stewardship theory assumes that stewards' interests are aligned with those of owners. This alignment reduces the need for extensive monitoring and control mechanisms. To understand the intricate dynamics of this relationship, some scholars have suggested reconciling these theories and applying them in a complementary and multiperspective manner (Le Breton-Miller et al., 2011).

According to stewardship theory, family business leaders act as stewards, motivated by higher-order needs (such as altruism) and a long-term orientation that surpasses that of nonfamily businesses. These family leaders identify with the organization and fully embrace its objectives, finding satisfaction in the company's success (Corbetta & Salvato, 2004b).

According to stewardship theory, the family is a valuable resource, and the importance of community values, mutual trust, and concern for others in family-run businesses is emphasized (Eddleston & Kellermanns, 2007). This theory posits that the collective culture within a family business promotes a strong commitment to achieving the corporate mission and ensuring its long-term stability, which contributes to its strategic flexibility (Zahra et al., 2008) and innovativeness (Craig & Dibrell, 2006). This culture helps business leaders nurture a group of loyal individuals who are essential for the success of their company. Since they are emotionally attached to their firm, these leaders invest everything they own in it, linking their fortunes and reputations to the business. As a result, a good steward becomes the ideal custodian of the company's assets, focused on pursuing the corporate mission and ensuring the growth of the business for future generations.

Stewardship theory highlights the need to enhance collaboration and decision-making processes between management and the board of directors by stimulating a sense of responsibility among managers. This sense of responsibility is derived from a deep understanding of the business, its scope, mission, and vision, which managers develop when they identify closely with the firm (Lane et al., 2006).

In the context of family firms, relationships grounded in trust and collaboration contribute to a governance system marked by a shared sense of responsibility toward multiple stakeholders, the promotion of prosocial behavior, and high levels of engagement across all organizational levels (Hernandez, 2012). This approach aligns with stewardship-oriented governance practices, which emphasize participative management (Eddleston & Kellermanns, 2007; Eddleston et al., 2012). Reflecting the relevance of this perspective, Neubaum et al. (2017) developed a stewardship climate scale to assess the extent to which the governance of a family business embodies stewardship principles.

The family plays a pivotal role in advancing sustainability within the firm. Marques et al. (2014) argue that strong family involvement serves as a key catalyst for greater engagement in CSR initiatives. Similarly, Déniz and Suárez (2005) highlight that CSR practices in family firms are often shaped by the family's ethical principles and cultural values. This orientation frequently translates into stronger stewardship of their internal and external stakeholder communities, such as employees and customers (Miller et al., 2008; Casprini et al., 2024).

4.4 Corporate Governance in Family Business

Corporate governance is a multifaceted discipline that is focused on the management and oversight of companies, aiming to ensure their effective and responsible governance. However, this governance varies widely across countries, reflecting diverse cultural, legal, institutional, and socioeconomic contexts. Specifically, the following three distinct models can be identified (Zattoni, 2020):

- Anglo–Saxon model: this model, which is commonly referred to as the market-oriented model and is prevalent in countries such as the United States and the United Kingdom, emphasizes a clear separation between ownership and control. It prioritizes shareholder interests and the maximization of shareholder value. Companies are supervised by a board of directors, primarily composed of independent directors who monitor managerial performance and protect shareholder interests. Key features include strong shareholder rights, a focus on financial performance, and additional oversight by auditors and audit committees.
- German model: this model, which is also known as the Rhineland model, is prevalent in Germany and other continental European countries. It integrates a wider range of stakeholders, including employees and suppliers, into the governance process. A dual-board structure is common in this model, with a supervisory board comprising both shareholder and employee representatives who oversee the management board. In contrast to the Anglo–Saxon model, smaller investors have less protection and access to information.
- Latin model: this model, which is adopted in many Latin American and Southern European countries, including Italy, is characterized by concentrated ownership and a powerful CEO who makes major decisions and represents the company externally. The board of directors mainly supervises and evaluates strategic decisions. Ownership and control are typically closely aligned, which can lead to a concentration of power, resulting in decisions that may not align with the interests of all stakeholders.

Over time, codes of good governance have been established to enhance corporate governance practices, protect stakeholders' interests and encourage companies to operate responsibly and with integrity by outlining ethical standards and principles. The first corporate governance code that was issued was the Cadbury Code in the UK in 1992. This code has significantly influenced corporate governance practices worldwide, laying the foundation for subsequent codes and regulations designed to increase corporate accountability and transparency. Its principles have been incorporated into various national and international governance frameworks, promoting better practices globally (e.g., OECD, 2023).

In the Italian context, the Corporate Governance Code (Borsa Italiana, 2020), that is, Codice di Autodisciplina, is issued by the Corporate Governance Committee established by Borsa Italiana, the Italian Stock Exchange. This code provides guidelines for listed companies in Italy, outlining best practices for corporate governance, and addresses key areas such as board composition, remuneration, internal controls, and shareholder relations.

In family businesses, the interplay among family dynamics, ownership, and management introduces intricate corporate governance challenges. These firms often struggle to balance familial relationships with business imperatives, which leads to unique governance dilemmas. Consequently, these firms often adopt corporate governance structures that differ significantly from those used in nonfamily businesses

(Bartholomeusz & Tanewski, 2006; Carney, 2005; Mustakallio et al., 2002; Siebels & zu Knyphausen-Aufseß, 2012). The close nature of family ties can impact decision-making processes, succession planning, and the alignment of interests between family members and nonfamily managers. Additionally, issues such as nepotism, family conflicts, and the integration of external governance mechanisms further complicate the governance landscape, driving the need for tailored strategies to ensure effective management.

The board of directors is a core element of the governance structure in family firms. Prior research has suggested that family firms are often hesitant to establish independent board structures because of concerns about losing decision-making discretion (Voordeckers et al., 2007). Independent directors, free from family pressures and not entangled in the family's social networks, can maintain the independence needed to effectively perform the board's control function (Klein et al., 2005). Conversely, insider-controlled boards tend to lead to entrenchment, as family members (i.e., insiders) have a strong incentive to exert control over the top management team (Long et al., 2005).

Traditionally, agency problems in companies can be mitigated through various governance mechanisms, including ownership concentration, boards of directors, executive compensation, debt, or dividends to reduce free cash flow, corporate takeovers, and legal investor protection. In family firms, additional mechanisms address agency issues between families and their representative shareholders. These include, for example, family assemblies, family councils, and family constitutions (Villalonga et al., 2015). The family assembly is composed of all family members, regardless of their role in the firm, provides information about the business and promotes a sense of belonging. Additionally, some companies have a family council, which collaborates with the board of directors to align family and shareholder goals, ensuring that family actions related to the firm are approved by the board. Another useful tool is the family constitution, a written agreement that establishes common values, policies, and rules among family members.

In family firms, executive compensation is another relevant issue. Research on this topic has produced mixed results, highlighting varying practices and outcomes related to the ways in which these firms compensate their executives. As noted by Gómez-Mejía et al. (2003), risk-averse CEOs might accept a lower salary in return for greater job security, especially when they have familial ties to owners. The authors found that the family connections of CEOs affect both the magnitude and composition of their compensation packages. CEOs who are family members in family-controlled firms earn less overall than external CEOs do, and this disparity grows as family ownership becomes more concentrated. On the other hand, Barontini and Bozzi (2011) demonstrated that Italian family firms offer higher salaries to their CEOs than do other firms do, with family CEOs earning more than nonfamily CEOs do. They view this as a form of rent extraction, suggesting that families overpay their CEOs to secure their loyalty and facilitate the expropriation of minority shareholders.

The effective management and planning of family succession is crucial in family businesses. Bertrand and Schoar (2006) highlighted that strong family ties can make it challenging for founders to separate family from business, often at the expense of company performance. As a result, the election of family CEOs is unlikely to be replaced by a system in which (nonfamily) management is challenged. However, research by Cucculelli and Micucci (2008) demonstrated that firms managed by heirs typically experience a more significant decline in performance post-succession than do those led by nonfamily CEOs.

Shared leadership is a governance structure that becoming more popular among family firms (Pearce & Conger, 2003). In fact, family firms increasingly implement shared leadership at the apex of their organizations (Arduino et al., 2022; Cater & Justis, 2010; Dennis et al., 2009; Frauenheim, 2009). The presence of two co-CEOs sharing power at the top of a company (dual leadership) is a particular case of shared leadership (Pearce & Sims, 2000). Co-CEOs can serve as a governance mechanism, with each CEO monitoring the other's behavior to prevent poor decisions, thereby reducing agency costs (Choi et al., 2018; Hasija et al., 2017). Additionally, having co-CEOs can broaden the range of leadership styles, skills, and competencies compared with having a single leader (Arnone & Stumpf, 2010). This arrangement enables a full utilization of coleaders' complementary knowledge and skills, stimulating a creative decision-making process (Cox et al., 2003). Finally, co-CEO arrangements effectively address the generational succession challenge in family businesses. As these firms grow and involve more family members, dividing the CEO role helps appease different family factions and guarantees a smoother leadership transition across generations (Alvarez et al., 2007).

Several key factors can support successful family transitions. These factors include comprehensive succession planning, clearly defined criteria for selecting a successor from within the family, the involvement of external consultants, and the adoption of suitable financial strategies (Cucculelli & Micucci, 2008). These managerial practices can play a significant role in ensuring smooth inheritance and the continued success of the family business to maintain the family legacy.

According to Martin (2001), certain values and practices need to be developed and continuously supported to ensure successful long-term family governance; these include the following:

– Culture and structure of open family communication: the development of a culture of open family communication can encourage transparency and trust and ensure that all family members are informed and engaged in decision-making processes.
– Valuing the overall family over individual or family branch needs: valuing the overall family needs promotes unity and shared goals, therefore reducing potential conflicts that may arise from prioritizing individuals over the family's collective interests.
– Importance of demonstrated competence in assigning responsibilities: roles need to be filled by those with the necessary skills and expertise to ensure more effective management.
– Effective generational succession plan for the survival of a family and its wealth: to ensure the long-term stability and prosperity of the family, it is critical to prepare future leaders and manage the transition of wealth and responsibilities.

- Creation of family conflict management processes: family firms should define structured methods for resolving potential disputes to maintain harmony and prevent conflicts from escalating.
- Creation and maintenance of an effective family governance plan: in family firms, establishing clear guidelines and systems for decision-making and oversight processes is crucial to facilitate the management and continuity of family affairs.

References

Aguilera, R. V., Williams, C. A., Conley, J. M., & Rupp, D. E. (2006). Corporate governance and social responsibility: A comparative analysis of the UK and the US. *Corporate Governance: An International Review, 14*(3), 147–158. https://doi.org/10.1111/j.1467-8683.2006.00495.x

Alvarez, J. L., Svejenova, S., & Vives, L. (2007). Leading in Pairs. *Sloan Management Review, 48*(4), 10–14.

Anderson, R. C., & Reeb, D. M. (2003). Founding-family ownership and firm performance: Evidence from the S&P 500. *The Journal of Finance, 58*(3), 1301–1328. https://doi.org/10.1111/1540-6261.00567

Ang, J. S., Cole, R. A., & Lin, J. W. (2000). Agency costs and ownership structure. *The Journal of Finance, 55*(1), 81–106. https://doi.org/10.1111/0022-1082.00201

Arduino, F. R., Zattoni, A., & Bozzolan, S. (2022). La diffusione dei modelli di leadership condivisa: Un'analisi esplorativa sulle società quotate italiane. *Corporate Governance and Research & Development Studies, 2*, 105–119. https://doi.org/10.3280/cgrds2-2021oa12542

Arnone, M., & Stumpf, S. A. (2010). Shared leadership: From rivals to co-CEOs. *Strategy & Leadership, 38*(2), 15–21. https://doi.org/10.1108/10878571011029019

Arregle, J. L., Hitt, M. A., Sirmon, D. G., & Very, P. (2007). The development of organizational social capital: Attributes of family firms. *Journal of Management Studies, 44*(1), 73–95. https://doi.org/10.1111/j.1467-6486.2007.00665.x

Astrachan, J. H., Klein, S. B., & Smyrnios, K. X. (2002). The F-PEC scale of family influence: A proposal for solving the family business definition problem. *Family Business Review, 15*(1), 45–58. https://doi.org/10.1111/j.1741-6248.2002.00045.x

Astrachan, J. H., & Shanker, M. C. (2003). Family businesses' contribution to the U.S. economy: A closer look. *Family Business Review, 16*(3), 211–219. https://doi.org/10.1177/08944865030160030601

AUB Observatory XV Edition. (2024, January 30). *Leadership change in Italian family businesses: Threat or opportunity?* Retrieved June 22, 2024, from https://www.aidaf.it/wp-content/uploads/2024/01/30/435-Presentazione-AUB-30.01.2024_formato_16_9.pdf

Barney, J. (1991). Firm resources and sustained competitive advantage. *Journal of Management, 17*(1), 99–120. https://doi.org/10.1177/014920639101700108

Barontini, R., & Bozzi, S. (2011). Board compensation and ownership structure: Empirical evidence for Italian listed companies. *Journal of Management & Governance, 15*(1), 59–89. https://doi.org/10.1007/s10997-009-9118-5

Barontini, R., & Caprio, L. (2006). The effect of family control on firm value and performance: Evidence from continental europe. *European Financial Management, 12*(5), 689–723. https://doi.org/10.1111/j.1468-036x.2006.00273.x

Barth, E., Gulbrandsen, T., & Schønea, P. (2005). Family ownership and productivity: The role of owner-management. *Journal of Corporate Finance, 11*(1–2), 107–127. https://doi.org/10.1016/j.jcorpfin.2004.02.001

Bartholomeusz, S., & Tanewski, G. A. (2006). The relationship between family firms and corporate governance. *Journal of Small Business Management, 44*(2), 245–267. https://doi.org/10.1111/j.1540-627x.2006.00166.x

Bennedsen, M., Nielsen, K. M., Perez-Gonzalez, F., & Wolfenzon, D. (2007). Inside the family firm: The role of families in succession decisions and performance. *The Quarterly Journal of Economics, 122*(2), 647–691. https://doi.org/10.1162/qjec.122.2.647

Berrone, P., Cruz, C., Gómez-Mejía, L. R., & Kintana, L. M. (2010). Socioemotional wealth and corporate responses to institutional pressures: Do family-controlled firms pollute less? *Administrative Science Quarterly, 55*(1), 82–113. https://doi.org/10.2189/asqu.2010.55.1.82

Berrone, P., Cruz, C., & Gomez-Mejia, L. R. (2012). Socioemotional wealth in family firms: Theoretical dimensions, assessments approaches and agenda for future research. *Family Business Review, 25*(3), 258–279. https://doi.org/10.1177/0894486511435355

Berrone, P., Gómez-Mejía, L. R., & Xu, K. (2023). The role of family ownership in norm-conforming environmental initiatives: Lessons from China. *Entrepreneurship Theory and Practice, 47*(5), 1915–1941. https://doi.org/10.1177/10422587221115362

Bertrand, M., & Schoar, A. (2006). The role of family in family firms. *Journal of Economic Perspectives, 20*(2), 73–96. https://doi.org/10.1257/jep.20.2.73

Borsa Italiana. (2020). *Corporate governance code.* Retrieved July 8, 2024, from https://www.borsaitaliana.it/comitato-corporate-governance/codice/2020.pdf

Cabrera-Suárez, M. K., Déniz-Déniz, M. C., & Martín-Santana, J. D. (2015). Family social capital, trust within the TMT, and the establishment of corporate goals related to nonfamily stakeholders. *Family Business Review, 28*(2), 145–162. https://doi.org/10.1177/0894486514526754

Carney, M. (2005). Corporate governance and competitive advantage in family-controlled firms. *Entrepreneurship Theory and Practice, 29*(3), 249–265. https://doi.org/10.1111/j.1540-6520.2005.00081.x

Casprini, E., Palumbo, R., & De Massis, A. (2024). Untangling the yarn: A contextualization of human resource management to the family firm setting. *Journal of Family Business Strategy, 15*(3), 100621. https://doi.org/10.1016/j.jfbs.2024.100621

Cater, J. J., & Justis, R. T. (2010). The development and implementation of shared leadership in multi-generational family firms. *Management Research Review, 33*(6), 563–585. https://doi.org/10.1108/01409171011050190

Cennamo, C., Berrone, P., Cruz, C., & Gomez-Mejia, L. R. (2012). Socioemotional wealth and proactive stakeholder engagement: Why family-controlled firms care more about their stakeholders. *Entrepreneurship Theory and Practice, 36*(6), 1153–1173. https://doi.org/10.1111/j.1540-6520.2012.00543.x

Choi, Y.-S., Hyeon, J., Jung, T., & Lee, W.-J. (2018). Audit pricing of shared leadership. *Emerging Markets Finance and Trade, 54*(2), 336–358. https://doi.org/10.1080/1540496x.2017.1348292

Chrisman, J. J., Chua, J. H., Kellermanns, F. W., & Chang, E. P. (2007). Are family managers agents or stewards? An exploratory study in privately held family firms. *Journal of Business Research, 60*(10), 1030–1038. https://doi.org/10.1016/j.jbusres.2006.12.011

Chrisman, J. J., Chua, J. H., & Litz, R. A. (2004). Comparing the agency costs of family and non-family firms: Conceptual issues and exploratory evidence. *Entrepreneurship Theory and Practice, 28*(4), 335–354. https://doi.org/10.1111/j.1540-6520.2004.00049.x

Chrisman, J. J., Chua, J. H., & Sharma, P. (2005). Trends and directions in the development of a strategic management theory of the family firm. *Entrepreneurship Theory and Practice, 29*(5), 555–575. https://doi.org/10.1111/j.1540-6520.2005.00098.x

Chrisman, J. J., & Patel, P. C. (2012). Variations in R&D investments of family and nonfamily firms: Behavioral agency and myopic loss aversion perspectives. *Academy of Management Journal, 55*(4), 976–997. https://www.jstor.org/stable/23317622

Claessens, S., Djankov, S., & Lang, L. H. P. (2000). The separation of ownership and control in East Asian Corporations. *Journal of Financial Economics, 58*(1–2), 81–112. https://doi.org/10.1016/s0304-405x(00)00067-2

References

Corbetta, G., & Salvato, C. (2004a). Self-serving or self-actualizing? Models of man and agency costs in different types of family firms: A commentary on "comparing the agency costs of family and non-family firms: Conceptual issues and exploratory evidence". *Entrepreneurship Theory and Practice, 28*(4), 355–362. https://doi.org/10.1111/j.1540-6520.2004.00050.x

Corbetta, G., & Salvato, C. A. (2004b). The board of directors in family firms: One size fits all? *Family Business Review, 17*(2), 119–134. https://doi.org/10.1111/j.1741-6248.2004.00008.x

Cox, J. F., Pearce, C. L., & Perry, M. L. (2003). Toward a model of shared leadership and distributed influence in the innovation process. In C. L. Pearce & J. A. Conger (Eds.), *Shared leadership: Reframing the hows and whys of leadership* (pp. 48–76). Sage.

Craig, J., & Dibrell, C. (2006). The natural environment, innovation, and firm performance: A comparative study. *Family Business Review, 19*(4), 275–288. https://doi.org/10.1111/j.1741-6248.2006.00075.x

Cucculelli, M., & Micucci, G. (2008). Family succession and firm performance: Evidence from Italian family firms. *Journal of Corporate Finance, 14*(1), 17–31. https://doi.org/10.1016/j.jcorpfin.2007.11.001

Davis, J. A. (2001). *Definitions and typologies of the family business*. Harvard Business School.

Davis, J. H., Schoorman, F. D., & Donaldson, L. (1997). Toward a stewardship theory of management. *Academy of Management Review, 22*(1), 20–47. https://doi.org/10.5465/amr.1997.9707180258

Deephouse, D. L., & Jaskiewicz, P. (2013). Do family firms have better reputations than non-family firms? An integration of socioemotional wealth and social identity theories. *Journal of Management Studies, 50*(3), 337–360. https://doi.org/10.1111/joms.12015

Déniz, M. D. L. C. D., & Suárez, M. K. C. (2005). Corporate social responsibility and family business in Spain. *Journal of Business Ethics, 56*(1), 27–41. https://link.springer.com/article/10.1007/s10551-004-3237-3

Dennis, S., Ramsey, D., & Turner, C. (2009). Dual or duel: Co-CEOs and firm performance. *Journal of Business & Economic Studies, 15*(1), 1–25.

Dick, M., Wagner, E., & Pernsteiner, H. (2021). Founder-controlled family firms, overconfidence, and corporate social responsibility engagement: Evidence from survey data. *Family Business Review, 34*(1), 71–92. https://doi.org/10.1177/08944865209187

Donaldson, L., & Davis, J. H. (1991). Stewardship theory or agency theory: CEO governance and shareholder returns. *Australian Journal of Management, 16*(1), 49–64. https://doi.org/10.1177/031289629101600103

Dyer, W. G., Jr., & Whetten, D. A. (2006). Family firms and social responsibility: Preliminary evidence from the S&P 500. *Entrepreneurship Theory and Practice, 30*(6), 785–802. https://doi.org/10.1111/j.1540-6520.2006.00151.x

Eddleston, K. A., & Kellermanns, F. W. (2007). Destructive and productive family relationships: A stewardship theory perspective. *Journal of Business Venturing, 22*(4), 545–565. https://doi.org/10.1016/j.jbusvent.2006.06.004

Eddleston, K. A., Kellermanns, F. W., & Zellweger, T. M. (2012). Exploring the entrepreneurial behavior of family firms: Does the stewardship perspective explain differences? *Entrepreneurship Theory and Practice, 36*(2), 347–367. https://doi.org/10.1111/j.1540-6520.2010.00402.x

Eddleston, K. A., & Kidwell, R. E. (2012). Parent–child relationships: Planting the seeds of deviant behavior in the family firm. *Entrepreneurship Theory and Practice, 36*(2), 369–386. https://doi.org/10.1111/j.1540-6520.2010.00403.x

El Ghoul, S., Guedhami, O., Wang, H., & Kwok, C. C. (2016). Family control and corporate social responsibility. *Journal of Banking & Finance, 73*, 131–146. https://doi.org/10.1016/j.jbankfin.2016.08.008

European Commission. (2009). *European Commission – family business*. Retrieved July 9, 2024 from https://single-market-economy.ec.europa.eu/smes/sme-fundamentals/family-business_en

EY and University of St. Gallen Family Business Index. (2023). Retrieved June 21, 2024, from https://familybusinessindex.com/

Fahlenbrach, R. (2009). Founder-CEOs, investment decisions, and stock market performance. *Journal of Financial and Quantitative Analysis, 44*(2), 439–466. https://doi.org/10.1017/s0022109009090139

Frauenheim, E. (2009). Co-CEOs: Two at the top. *Workforce Management, 88*(6), 40.

Gómez-Mejía, L. R., Cruz, C., Berrone, P., & De Castro, J. (2011). The bind that ties: Socioemotional wealth preservation in family firms. *Academy of Management Annals, 5*(1), 653–707. https://doi.org/10.5465/19416520.2011.593320

Gómez-Mejía, L. R., Haynes, K. T., Núñez-Nickel, M., Jacobson, K. J. L., & Moyano-Fuentes, J. (2007). Socioemotional wealth and business risks in family-controlled firms: Evidence from Spanish olive oil mills. *Administrative Science Quarterly, 52*(1), 106–137. https://doi.org/10.2189/asqu.52.1.106

Gómez-Mejía, L. R., Larraza-Kintana, M., & Makri, M. (2003). The determinants of executive compensation in family-controlled public corporations. *Academy of Management Journal, 46*(2), 226–237. https://doi.org/10.5465/30040616

Gómez-Mejía, L. R., Makri, M., & Kintana, M. L. (2010). Diversification decisions in family-controlled firms. *Journal of Management Studies, 47*(2), 223–252. https://doi.org/10.1111/j.1467-6486.2009.00889.x

Gómez-Mejía, L. R., Muñoz-Bullón, F., Requejo, I., & Sanchez-Bueno, M. J. (2025). Ethical correlates of family control: Socioemotional wealth, environmental performance, and financial returns. *Journal of Business Ethics, 198*, 893–917. https://doi.org/10.1007/s10551-025-05943-9

Gómez-Mejía, L. R., Nunez-Nickel, M., & Gutierrez, I. (2001). The role of family ties in agency contracts. *Academy of Management Journal, 44*(1), 81–95. https://doi.org/10.5465/3069338

Habbershon, T. G., & Williams, M. L. (1999). A resource-based framework for assessing the strategic advantages of family firms. *Family Business Review, 12*(1), 1–25. https://doi.org/10.1111/j.1741-6248.1999.00001.x

Hasija, D. B., Ellstrand, A. E., Worrell, D. L., & Dixon-Fowler, H. (2017). Two heads may be more responsible than one: Co-CEOs and corporate social performance. *Journal of Management Policy and Practice, 18*(2), 9–21.

Hernandez, M. (2012). Toward an understanding of the psychology of stewardship. *Academy of Management Review, 37*(2), 172–193. https://doi.org/10.5465/amr.2010.0363

Holderness, C. G., & Sheehan, D. P. (1988). The role of majority shareholders in publicly held corporations. *Journal of Financial Economics, 20*, 317–346. https://doi.org/10.1016/0304-405x(88)90049-9

Huang, Y. C., Yang, M. L., & Wong, Y. J. (2016). The effect of internal factors and family influence on firms' adoption of green product innovation. *Management Research Review, 39*(10), 1167–1198. https://doi.org/10.1108/MRR-02-2015-0031

ISTAT. (2023, November 14). *2023 Permanent census of enterprises: Initial results.* Retrieved June 23, 2024, from https://www.istat.it/it/files/2023/11/REPORTCensimprese.pdf

Jaskiewicz, P., Uhlenbruck, K., Balkin, D. B., & Reay, T. (2013). Is nepotism good or bad? Types of nepotism and implications for knowledge management. *Family Business Review, 26*(2), 121–139. https://doi.org/10.1177/0894486512470841

Jensen, M. C., & Meckling, W. H. (1976). Theory of the firm: Managerial behavior, agency costs and ownership structure. *Journal of Financial Economics, 3*(4), 305–360. https://doi.org/10.1016/0304-405x(76)90026-x

Kellermanns, F. W., Eddleston, K. A., & Zellweger, T. (2012). Extending the socioemotional wealth perspective: A look at the dark side. *Entrepreneurship Theory and Practice, 36*(6), 1175–1182. https://doi.org/10.1111/j.1540-6520.2012.00544.x

Klein, P., Shapiro, D., & Young, J. (2005). Corporate governance, family ownership and firm value: The Canadian evidence. *Corporate Governance: An International Review, 13*(6), 769–784. https://doi.org/10.1111/j.1467-8683.2005.00469.x

Labelle, R., Hafsi, T., Francoeur, C., & Ben Amar, W. (2018). Family firms' corporate social performance: A calculated quest for socioemotional wealth. *Journal of Business Ethics, 148*(3), 511–525. https://doi.org/10.1007/s10551-015-2982-9

References

Lane, S., Astrachan, J., Keyt, A., & McMillan, K. (2006). Guidelines for family business boards of directors. *Family Business Review, 19*(2), 147–167. https://doi.org/10.1111/j.1741-6248.2006.00052.x

La Porta, R., Lopez-De-Silanes, F., & Shleifer, A. (1999). Corporate ownership around the world. *The Journal of Finance, 54*(2), 471–517. https://doi.org/10.1111/0022-1082.00115

Le Breton-Miller, I., & Miller, D. (2013). Socioemotional wealth across the family firm life cycle: A commentary on "family business survival and the role of boards". *Entrepreneurship Theory and Practice, 37*(6), 1391–1397. https://doi.org/10.1111/etap.12072

Le Breton-Miller, I., & Miller, D. (2016). Family firms and practices of sustainability: A contingency view. *Journal of Family Business Strategy, 7*(1), 26–33. https://doi.org/10.1016/j.jfbs.2015.09.001

Le Breton-Miller, I., Miller, D., & Lester, R. H. (2011). Stewardship or agency? A social embeddedness reconciliation of conduct and performance in public family businesses. *Organization Science, 22*(3), 704–721. https://doi.org/10.1287/orsc.1100.0541

Long, T., Dulewicz, V., & Gay, K. (2005). The role of the non-executive director: Findings of an empirical investigation into the differences between listed and unlisted UK boards. *Corporate Governance: An International Review, 13*(5), 667–679. https://doi.org/10.1111/j.1467-8683.2005.00458.x

Lumpkin, G. T., & Bacq, S. (2022). Family business, community embeddedness, and civic wealth creation. *Journal of Family Business Strategy, 13*(2), 100469. https://doi.org/10.1016/j.jfbs.2021.100469

Machek, O. (2016). The development of family business literature in 2000–2014: What can we learn from Scopus. *International Journal of Economics and Statistics, 4*.

Marques, P., Presas, P., & Simon, A. (2014). The heterogeneity of family firms in CSR engagement: The role of values. *Family Business Review, 27*(3), 206–227. https://doi.org/10.1177/0894486514539004

Martin, H. F. (2001). Is family governance an oxymoron? *Family Business Review, 14*(2), 91–96. https://doi.org/10.1111/j.1741-6248.2001.00091.x

Maury, B. (2006). Family ownership and firm performance: Empirical evidence from Western European corporations. *Journal of Corporate Finance, 12*(2), 321–341. https://doi.org/10.1016/j.jcorpfin.2005.02.002

Miller, D., Le Breton-Miller, I., & Lester, R. H. (2013). Family firm governance, strategic conformity, and performance: Institutional vs. strategic perspectives. *Organization Science, 24*(1), 189–209. https://doi.org/10.1287/orsc.1110.0728

Miller, D., Le Breton-Miller, I., Lester, R. H., & Cannella, A. A. (2007). Are family firms really superior performers? *Journal of Corporate Finance, 13*(5), 829–858. https://doi.org/10.1016/j.jcorpfin.2007.03.004

Miller, D., Le Breton-Miller, I., & Scholnick, B. (2008). Stewardship vs. stagnation: An empirical comparison of small family and non-family businesses. *Journal of Management Studies, 45*(1), 51–78. https://doi.org/10.1111/j.1467-6486.2007.00718.x

Minichilli, A., Nordqvist, M., Corbetta, G., & Amore, M. D. (2014). CEO succession mechanisms, organizational context, and performance: A socio-emotional wealth perspective on family-controlled firms. *Journal of Management Studies, 51*(7), 1153–1179. https://doi.org/10.1111/joms.12095

Morck, R., Shleifer, A., & Vishny, R. W. (1988). Management ownership and market valuation: An empirical analysis. *Journal of Financial Economics, 20*, 293–315. https://doi.org/10.1016/0304-405x(88)90048-7

Morck, R., & Yeung, B. (2003). Agency problems in large family business groups. *Entrepreneurship Theory and Practice, 27*(4), 367–382. https://doi.org/10.1111/1540-8520.t01-1-00015

Morck, R., & Yeung, B. (2004). Family control and the rent-seeking society. *Entrepreneurship Theory and Practice, 28*(4), 391–409. https://doi.org/10.1111/j.1540-6520.2004.00053.x

Mustakallio, M., Autio, E., & Zahra, S. A. (2002). Relational and contractual governance in family firms: Effects on strategic decision making. *Family Business Review, 15*(3), 205–222. https://doi.org/10.1111/j.1741-6248.2002.00205.x

Neubaum, D. O., Thomas, C. H., Dibrell, C., & Craig, J. B. (2017). Stewardship climate scale: An assessment of reliability and validity. *Family Business Review, 30*(1), 37–60. https://doi.org/10.1177/0894486516673701

OECD. (2023). *G20/OECD principles of corporate governance 2023*. OECD Publishing.

Pearce, C. L., & Conger, J. A. (2003). All those years ago: The historical underpinnings of shared leadership. In Shared leadership: Reframing the hows and whys of leadership (pp. 1–18). SAGE Publications, Inc.

Pearce, C. L., & Sims, H. P. (2000). Shared leadership: Toward a multi-level theory of leadership. In M. Beyerlein, D. Johnson, & S. Beyerlein (Eds.), *Advances in interdisciplinary studies of work teams* (pp. 115–139). JAI Press.

Pearson, A. W., Carr, J. C., & Shaw, J. C. (2008). Toward a theory of familiness: A social capital perspective. *Entrepreneurship Theory and Practice, 32*(6), 949–969. https://doi.org/10.1111/j.1540-6520.2008.00265.x

Rau, S. B. (2014). Resource-based view of family firms. In L. Melin, M. Nordqvist, & P. Sharma (Eds.), *The SAGE handbook of family business* (pp. 321–339). Sage.

Rees, W., & Rodionova, T. (2015). The influence of family ownership on corporate social responsibility: An international analysis of publicly listed companies. *Corporate Governance: An International Review, 23*(3), 184–202. https://doi.org/10.1111/corg.12086

Sanchez-Ruiz, P., Daspit, J. J., Holt, D. T., & Rutherford, M. W. (2019). Family social capital in the family firm: A taxonomic classification, relationships with outcomes, and directions for advancement. *Family Business Review, 32*(2), 131–153. https://doi.org/10.1177/0894486519836833

Schulze, W. S., Lubatkin, M. H., & Dino, R. N. (2003). Toward a theory of agency and altruism in family firms. *Journal of Business Venturing, 18*(4), 473–490. https://doi.org/10.1016/s0883-9026(03)00054-5

Schulze, W. S., Lubatkin, M. H., Dino, R. N., & Buchholtz, A. K. (2001). Agency relationships in family firms: Theory and evidence. *Organization Science, 12*(2), 99–116. https://doi.org/10.1287/orsc.12.2.99.10114

Sciascia, S., Mazzola, P., & Kellermanns, F. W. (2014). Family management and profitability in private family-owned firms: Introducing generational stage and the socioemotional wealth perspective. *Journal of Family Business Strategy, 5*(2), 131–137. https://doi.org/10.1016/j.jfbs.2014.03.001

Siebels, J. F., & zu Knyphausen-Aufseß, D. (2012). A review of theory in family business research: The implications for corporate governance. *International Journal of Management Reviews, 14*(3), 280–304. https://doi.org/10.1111/j.1468-2370.2011.00317.x

Sirmon, D. G., & Hitt, M. A. (2003). Managing resources: Linking unique resources, management, and wealth creation in family firms. *Entrepreneurship Theory and Practice, 27*(4), 339–358. https://doi.org/10.1111/1540-8520.t01-1-00013

Stock, C., Pütz, L., Schell, S., & Werner, A. (2024). Corporate social responsibility in family firms: Status and future directions of a research field. *Journal of Business Ethics, 190*(1), 199–259. https://doi.org/10.1007/s10551-023-05382-4

Tagiuri, R., & Davis, J. (1996). Bivalent attributes of the family firm. *Family Business Review, 9*(2), 199–208. https://doi.org/10.1111/j.1741-6248.1996.00199.x

Veiga, P. M. (2025). Key drivers of green innovation in family firms: a machine learning approach. *Journal of Family Business Management, 15*(2), 346–372. https://doi.org/10.1108/JFBM-08-2024-0191

Villalonga, B., & Amit, R. (2006). How do family ownership, control and management affect firm value? *Journal of Financial Economics, 80*(2), 385–417. https://doi.org/10.1016/j.jfineco.2004.12.005

Villalonga, B., Amit, R., Trujillo, M.-A., & Guzmán, A. (2015). Governance of family firms. *Annual Review of Financial Economics, 7*(1), 635–654. https://doi.org/10.1146/annurev-financial-110613-034357

References

Voordeckers, W., Van Gils, A., & Van Den Heuvel, J. (2007). Board composition in small and medium-sized family firms. *Journal of Small Business Management, 45*(1), 137–156. https://doi.org/10.1111/j.1540-627x.2007.00204.x

Ward, J. L. (2004). *Perpetuating the family business*. Palgrave Macmillan.

Yang, J., & Basile, K. (2022). Communicating corporate social responsibility: External stakeholder involvement, productivity and firm performance. *Journal of Business Ethics, 178*(2), 501–517. https://doi.org/10.1007/s10551-021-04812-5

Zahra, S. A., Hayton, J. C., Neubaum, D. O., Dibrell, C., & Craig, J. (2008). Culture of family commitment and strategic flexibility: The moderating effect of stewardship. *Entrepreneurship Theory and Practice, 32*(6), 1035–1054. https://doi.org/10.1111/j.1540-6520.2008.00271.x

Zattoni, A. (2020). *Corporate governance. How to design good companies*. Bocconi University Press.

Zientara, P. (2017). Socioemotional wealth and corporate social responsibility: A critical analysis. *Journal of Business Ethics, 144*(1), 185–199. https://doi.org/10.1007/s10551-015-2848-1

Chapter 5
Empirical Analysis of Italian Listed Family Businesses

5.1 Objectives and Scope of the Study

This study aims to empirically analyze the key characteristics of family businesses within the Italian context, specifically focusing on their unique governance structures and sustainability practices. By investigating a selected sample of domestically listed family firms, this research seeks to offer valuable insights into how these businesses are responding to the rising demand for sustainable governance, particularly in light of recent European directives.

After the data sources and the variables being examined are introduced, the characteristics of the selected sample of Italian domestically listed family firms are outlined. The findings of a descriptive analysis are subsequently presented and discussed. Specifically, the study includes a comparative analysis of governance and sustainability features, highlighting best practices and lessons learned from the empirical data.

5.2 Methodology and Data Sources

A variety of data collection sources were utilized to conduct the analysis. The resulting dataset includes several types of data, as follows:

- Governance data: data related to corporate governance were hand-collected from the annual Corporate Governance Reports available on the Borsa Italiana (the Italian Stock Exchange) website (Borsa Italiana, 2024), as well as from the LSEG Workspace (formerly known as the Thomson Reuters' Refinitiv Database). The LSEG workspace is an internationally recognized data provider that offers extensive governance information, such as board composition, enabling the

assessment of companies' governance practices and compliance with corporate governance codes' recommendations (LSEG Workspace, 2024).
- Ownership data: data related to ownership were collected using the information available on the CONSOB (Italian Companies and Stock Exchange Commission) website. CONSOB is the regulatory authority for the securities market in Italy, and it maintains various databases that include information on listed Italian companies, including their ownership structures, shareholder compositions, and significant shareholdings (CONSOB, 2024).
- Sustainability data: data related to sustainability were gathered from the LSEG workspace (formerly known as the Thomson Reuters' Refinitiv Database). This platform offers a comprehensive suite of ESG data and sustainability metrics. The LSEG workspace consolidates different data sources, providing insights into company performance related to environmental practices, social responsibility, and governance structures (LSEG Workspace, 2024).
- Firm-level data: data related to the industries to which the companies of the sample belong were gathered from LSEG Workspace (LSEG Workspace, 2024), whereas data related to the areas of Italy where the companies are located were collected from using the information available on the Registro Imprese website, which is the main public registry of Italian companies, providing data from Chambers of Commerce (Registro Imprese, 2024).

Tables 5.1, 5.2, 5.3, 5.4, 5.5 provide specific descriptions of the variables included in this study, along with their respective data sources.

Table 5.1 Variables description and data sources: Governance

Variable	Description	Data source
Board size	The total number of board members at the end of the fiscal year.	LSEG workspace, Borsa Italiana
Independent board members	Percentage of independent board members as reported by the company.	LSEG workspace, Borsa Italiana
Executive board members	Percentage of executive board members.	LSEG workspace, Borsa Italiana
CEO duality	Dummy variable equal to 1 if the CEO simultaneously chair the board, and 0 otherwise.	LSEG workspace, Borsa Italiana
Board gender diversity	Percentage of female board members.	LSEG workspace
Average board tenure	Average number of years each board member has been on the board.	LSEG workspace
Board member affiliations	Average number of other corporate affiliations for the board member.	LSEG workspace
Co-CEO	Dummy variable equal to 1 if there are co-CEOs, and 0 otherwise.	Borsa Italiana
Nomination board committee	Dummy variable equal to 1 if the company has a nomination board committee, and 0 otherwise.	LSEG workspace

(continued)

5.3 Sample

Table 5.1 (continued)

Variable	Description	Data source
Corporate governance board committee	Dummy variable equal to 1 if the company has a corporate governance board committee, and 0 otherwise.	LSEG workspace
Compensation board committee	Dummy variable equal to 1 if the company has a compensation board committee, and 0 otherwise.	LSEG workspace
Audit board committee	Dummy variable equal to 1 if the company has an audit board committee, and 0 otherwise.	LSEG workspace
CSR sustainability committee	Dummy variable equal to 1 if the company has a CSR sustainability committee, and 0 otherwise.	LSEG workspace

Notes: CEO denotes Chief Executive Officer; CSR denotes Corporate Social Responsibility
Source: Author's elaboration on Borsa Italiana and LSEG workspace (Accessed July 21, 2024)

Table 5.2 Variable description and data sources: Sustainability

Variable	Description	Data source
ESG score	ESG Score is an overall company score based on the self-reported information in the environmental, social, and corporate governance pillars.	LSEG Workspace
Environmental pillar score	Environment Pillar Score is the weighted average relative rating of a company based on the reported environmental information and the resulting three environmental category scores.	LSEG Workspace
Social pillar score	Social Pillar Score is the weighted average relative rating of a company based on the reported social information and the resulting four social category scores.	LSEG Workspace
Governance pillar score	Governance Pillar Score is the weighted average relative rating of a company based on the reported governance information and the resulting three governance category scores.	LSEG Workspace
Resource use score	Resource use category score reflects a company's performance and capacity to reduce the use of materials, energy or water, and to find more eco-efficient solutions by improving supply chain management.	LSEG Workspace
Emissions score	Emission category score measures a company's commitment and effectiveness toward reducing environmental emission in the production and operational processes.	LSEG Workspace
Environmental innovation score	Environmental innovation category score reflects a company's capacity to reduce the environmental costs and burdens for its customers, and thereby creating new market opportunities through new environmental technologies and processes or eco-designed products.	LSEG Workspace
Workforce score	Workforce category score measures a company's effectiveness toward job satisfaction, healthy and safe workplace, maintaining diversity and equal opportunities, and development opportunities for its workforce.	LSEG Workspace
Human rights score	Human rights category score measures a company's effectiveness toward respecting the fundamental human rights conventions.	LSEG Workspace

(continued)

Table 5.2 (continued)

Variable	Description	Data source
Community score	Community category score measures the company's commitment toward being a good citizen, protecting public health and respecting business ethics.	LSEG Workspace
Product responsibility score	Product responsibility category score reflects a company's capacity to produce quality goods and services integrating the customer's health and safety, integrity and data privacy.	LSEG Workspace
Management score	Management category score measures a company's commitment and effectiveness toward following best practice corporate governance principles.	LSEG Workspace
Shareholders score	Shareholders category score measures a company's effectiveness toward equal treatment of shareholders and the use of anti-takeover devices.	LSEG Workspace
CSR Strategy score	CSR strategy category score reflects a company's practices to communicate that it integrates the economic (financial), social and environmental dimensions into its day-to-day decision-making processes.	LSEG Workspace

Notes: CSR denotes Corporate Social Responsibility; ESG denotes environmental, social and governance
Source: Author's elaboration on LSEG Workspace (Accessed July 21, 2024)

Table 5.3 Descriptions of variables and data sources: Sustainable Development Goals

Variable	Description	Data source
SDG 1: No poverty	Does the company support the UN Sustainable Development Goal 1 (SDG 1) No Poverty? Dummy variable equal to 1 if the company is supporting Goal 1 of SDG to End poverty in all its forms everywhere, and 0 otherwise.	LSEG workspace
SDG 2: Zero hunger	Does the company support the UN Sustainable Development Goal 2 (SDG 2) Zero Hunger? Dummy variable equal to 1 if the company is supporting Goal 2 of SDG to end hunger, achieve food security and improved nutrition, and promote sustainable agriculture, and 0 otherwise.	LSEG workspace
SDG 3: Good health and well-being	Does the company support the UN Sustainable Development Goal 3 (SDG 3) Good Health and Well-being? Dummy variable equal to 1 if the company is supporting Goal 3 of SDG to ensure healthy lives and promote well-being for all, and 0 otherwise.	LSEG workspace
SDG 4: Quality education	Does the company support the UN Sustainable Development Goal 4 (SDG 4) Quality Education? Dummy variable equal to 1 if the company is supporting Goal 4 of SDG to ensure inclusive and equitable quality education and promote lifelong learning opportunities for all, and 0 otherwise.	LSEG workspace
SDG 5: Gender equality	Does the company support the UN Sustainable Development Goal 5 (SDG 5) Gender Equality? Dummy variable equal to 1 if the company is supporting Goal 5 of SDG to achieve gender equality and empower all women and girls, and 0 otherwise.	LSEG workspace

(continued)

5.4 Comparative Analysis of Governance and Sustainability Practices

Table 5.3 (continued)

Variable	Description	Data source
SDG 6: Clean water and sanitation	Does the company support the UN Sustainable Development Goal 6 (SDG 6) Clean Water and Sanitation? Dummy variable equal to 1 if the company is supporting Goal 6 of SDG to ensure access to water and sanitation for all, and 0 otherwise.	LSEG workspace
SDG 7: Affordable and clean energy	Does the company support the UN Sustainable Development Goal 7 (SDG 7) Affordable and Clean Energy? Dummy variable equal to 1 if the company is supporting Goal 7 of SDG to ensure access to affordable, reliable, sustainable and modern energy for all, and 0 otherwise.	LSEG workspace
SDG 8: Decent work and economic growth	Does the company support the UN Sustainable Development Goal 8 (SDG 8) Decent Work and Economic Growth? Dummy variable equal to 1 if the company is supporting Goal 8 of SDG to promote inclusive and sustainable economic growth, employment, and decent work for all, and 0 otherwise.	LSEG workspace
SDG 9: Industry, innovation and infrastructure	Does the company support the UN Sustainable Development Goal 9 (SDG 9) Industry, Innovation, and Infrastructure? Dummy variable equal to 1 if the company is supporting Goal 9 of SDG to build resilient infrastructure, promote sustainable industrialization and foster innovation, and 0 otherwise.	LSEG workspace
SDG 10: Reduced inequality	Does the company support the UN Sustainable Development Goal 10 (SDG 10) Reduced Inequality? Dummy variable equal to 1 if the company is supporting Goal 10 of SDG to reduce inequality within and among countries, and 0 otherwise.	LSEG workspace
SDG 11: Sustainable cities and communities	Does the company support the UN Sustainable Development Goal 11 (SDG 11) Sustainable Cities and Communities? Dummy variable equal to 1 if the company is supporting Goal 11 of SDG to make cities inclusive, safe, resilient and sustainable, and 0 otherwise.	LSEG workspace
SDG 12: Responsible consumption and production	Does the company support the UN Sustainable Development Goal 12 (SDG 12) Responsible Consumption and Production? Dummy variable equal to 1 if the company is supporting Goal 12 of SDG to ensure sustainable consumption and production patterns, and 0 otherwise.	LSEG workspace
SDG 13: Climate action	Does the company support the UN Sustainable Development Goal 13 (SDG 13) Climate Action? Dummy variable equal to 1 if the company is supporting Goal 13 of SDG to take urgent action to combat climate change and its impacts, and 0 otherwise.	LSEG workspace
SDG 14: Life below water	Does the company support the UN Sustainable Development Goal 14 (SDG 14) Life Below Water? Dummy variable equal to 1 if the company is supporting Goal 14 of SDG to conserve and sustainably use of the oceans, seas and marine resources, and 0 otherwise.	LSEG workspace

(continued)

Table 5.3 (continued)

Variable	Description	Data source
SDG 15: Life on land	Does the company support the UN Sustainable Development Goal 15 (SDG 15) Life on Land? Dummy variable equal to 1 if the company is supporting Goal 15 of SDG to sustainably manage forests, combat desertification, halt and reverse land degradation, halt biodiversity loss, and 0 otherwise.	LSEG workspace
SDG 16: Peace and justice strong institutions	Does the company support the UN Sustainable Development Goal 16 (SDG 16) Peace and Justice Strong Institutions? Dummy variable equal to 1 if the company is supporting Goal 16 of SDG to promote just, peaceful and inclusive societies, and 0 otherwise.	LSEG workspace
SDG 17: Partnerships to achieve the goal	Does the company support the UN Sustainable Development Goal 17 (SDG 17) Partnerships to achieve the Goal? Dummy variable equal to 1 if the company is supporting Goal 17 of SDG to revitalize the global partnership for sustainable development, and 0 otherwise.	LSEG workspace

Notes: SDG denotes Sustainable Development Goal; UN denotes United Nations
Source: Author's elaboration on LSEG workspace (Accessed July 21, 2024)

Table 5.4 Description of variables and data sources: Stakeholder engagement and corporate policies related to sustainability

Variable	Description	Data source
Stakeholder Engagement	Dummy variable equal to 1 if the company explains how it engages with its stakeholders, in particular by: – Providing information on how the company is engaging with its stakeholders, how it is involving the stakeholders in its decision-making process; what procedures are in place for engagement. – Focusing on having established two-way communication between the company and its various stakeholders.	LSEG Workspace
Policy Emissions	Dummy variable equal to 1 if the company has a policy to improve emission reduction, as follows: – In scope are the various forms of emissions to land, air or water from the company's core activities. – Processes, mechanisms, or programs in place as to what the company is doing to reduce emissions in its operations. – System or a set of formal, documented processes for controlling emissions and driving continuous improvement.	LSEG Workspace
Policy Environmental Supply Chain	Dummy variable equal to 1 if the company has a policy to include its supply chain in the company's efforts to lessen its overall environmental impact. It considers: – Legal compliance data on the supply chain to reduce environmental impact is in scope. – Data on collaboration with suppliers toward reducing their environmental impacts. – Data on the reduction of environmental impacts at the suppliers' operations.	LSEG Workspace

(continued)

5.4 Comparative Analysis of Governance and Sustainability Practices

Table 5.4 (continued)

Variable	Description	Data source
Policy Energy Efficiency	Dummy variable equal to 1 if the company has a policy to improve its energy efficiency, including: – In scope are the various forms of processes/mechanisms/procedures to improve energy use in operation efficiently. – System or a set of formal documented processes for efficient use of energy and driving continuous improvement.	LSEG Workspace
Policy Human Rights	Dummy variable equal to 1 if the company has a policy to ensure the respect of human rights in general, including: – Information to be on ensuring the respect of human rights. – Consider a process on general fundamental human rights.	LSEG Workspace
Policy Forced Labor	Dummy variable equal to 1 if the company has a policy to avoid the use of forced labor, including: – Actions, programs or initiatives to avoid forced or compulsory labor for the company or its suppliers. – Practices to avoid any work for which people are forced to do against their will. – Consider information from industry code such as the Electronic Industry Citizenship Coalition (EICC) code of conduct and Pharmaceutical Industry Principles (PSCI). – Legal compliance data are considered.	LSEG Workspace
Policy Employee Health & Safety	Dummy variable equal to 1 if the company has a policy to improve employee health & safety. It considers: – Processes or initiatives in place to reduce occupational accidents, injuries, illness for employees of the company. – Information may refer to a system, project or a set of formal, documented processes for controlling health and safety impacts and driving continuous improvement. – Consider the process to reduce commuting accidents.	LSEG Workspace
Policy Diversity and Opportunity	Dummy variable equal to 1 if the company has a policy to drive diversity and equal opportunity, including: – Program or practice to promote diversity and equal opportunities within the workforce. – Information on the promotion of women, minorities, disabled employees, or employment from any age, ethnicity, race, nationality, and religion. – Information from the code of conduct mentioning diversity policy together with the reporting of violations.	LSEG Workspace
Policy Shareholder Engagement	Dummy variable equal to 1 if the company has a policy to facilitate shareholder engagement, resolutions, or proposals, including: – In scope, are the data on company facilitating shareholders to have the right to ask a question to the board or management. – Allowing shareholders to table resolutions or shareholder proposals at the shareholder meetings.	LSEG Workspace
Policy Executive Compensation ESG Performance	Dummy variable equal to 1 if the company has an extrafinancial performance-oriented compensation policy. – The compensation policy includes remuneration for the CEO, executive directors, nonboard executives, and other management bodies based on ESG or sustainability factors.	LSEG Workspace

(continued)

Table 5.4 (continued)

Variable	Description	Data source
Policy Community Involvement	Dummy variable equal to 1 if the company has a policy to improve its good corporate citizenship. It considers: – Involvement in the community through donations, volunteering, philanthropic activities, and community investments. – Involvement in corporate social responsibility programs in education, health, and the environment.	LSEG Workspace
Policy Business Ethics	Dummy variable equal to 1 if the company describes in the code of conduct that it strives to maintain the highest level of general business ethics. It considers: – Information on respecting general business ethics or integrity. – Information from the code of conduct section.	LSEG Workspace
Policy Bribery and Corruption	Dummy variable equal to 1 if the company describes in the code of conduct that it strives to avoid bribery and corruption at all its operations. It considers: – Policy in the code of conduct against the bribery and corruption in its operations. – Information from the code of conduct section in any report. – Inappropriate/improper payment, special favors, extortion or kickback. – Legal compliance data are not considered.	LSEG Workspace

Notes: CEO denotes Chief Operating Officer; ESG denotes environmental, social, and governance
Source: Author's elaboration on LSEG Workspace (Accessed July 21, 2024)

Table 5.5 Variable description and data sources: Ownership, industry, location

Variable	Description	Data source
Family ownership	Dummy variable equal to 1 if the family owns more than 25% of the firm's capital, and 0 otherwise.	CONSOB
Industry	Dummy variable equal to 1 if the company belongs to one of the following industries: (i) Basic materials. (ii) Consumer cyclicals. (iii) Consumer non-cyclicals. (iv) Energy. (v) Healthcare. (vi) Industrials. (vii) Technology. (viii) Utilities.	LSEG workspace
Location	Dummy variable equal to 1 if the company belongs to one of the following areas of Italy: (i) Center. (ii) North East. (iii) North West. (iv) South and Islands.	Registro Imprese

Source: Author's elaboration on CONSOB, Registro Imprese, and LSEG workspace (Accessed July 21, 2024)

5.3 Sample

In this study, the sample consists of a selection of domestic family firms listed on the Milan Stock Exchange from January 1, 2017, to December 31, 2022. This time frame was chosen so that the influence of the NFRD, which was formally introduced in 2014 following approval by the EU, on the companies being examined could be assessed. The NFRD, which is also known as Directive 2014/95/EU, requires large public interest entities with more than 500 employees to disclose nonfinancial information, including ESG factors (EU Non-financial Reporting Directive, 2014). The NFRD was implemented in Italian law in 2016 through Legislative Decree No. 254 (Gazzetta Ufficiale della Repubblica Italiana, 2016), which mandated that applicable companies begin to report their nonfinancial information for the financial year starting on or after January 1, 2017. This was implemented to increase transparency and provide investors and other stakeholders with essential information on the sustainability efforts of companies.

To ensure a representative and robust sample, we established the following strict inclusion criteria:

- Only firms operating under a traditional model were included;
- Firms within the financial sector were explicitly excluded because they are subject to sector-specific regulations that affect their governance structure;
- Foreign companies listed on the Milan Stock Exchange were excluded, as they are subject to different corporate law regulations.

For the purposes of our study, we use the definition of a family business on the basis of control. Consistent with previous research (e.g., Anderson & Reeb, 2003; Barth et al., 2005), we define family control as the proportion of equity owned by family members that enables them to exert control over the company. Specifically, in this study, a business is considered family-controlled if a family owns 25% of the shares, in line with prior research (Amore et al., 2011).

After excluding firms with missing data, we obtained an unbalanced sample consisting of 140 domestic family firms listed on the Milan Stock Exchange between 2017 and 2022. Of these, 83 firms are required to disclose nonfinancial information, following the NFRD.

As shown in Fig. 5.1, the distribution of the sample by industry indicates that companies belong primarily to the consumer cyclical sector (i.e., 38%) and the industrial sector (i.e., 23%), followed by technology (i.e., 12%), basic materials (i.e., 8%), consumer non-cyclicals (i.e., 7%), and healthcare (i.e., 6%). Finally, there are companies in the real estate and utilities sectors (i.e., 3% each).

As shown in Fig. 5.2, the geographical distribution of the sample shows that the companies are predominantly located in the northern regions of Italy, particularly in the Northwest (i.e., 52%) and Northeast (i.e., 22%). This is followed by companies in the Center of Italy (i.e., 23%), whereas only a very small fraction of the sample is situated in the South of Italy or on the Islands (i.e., 3%).

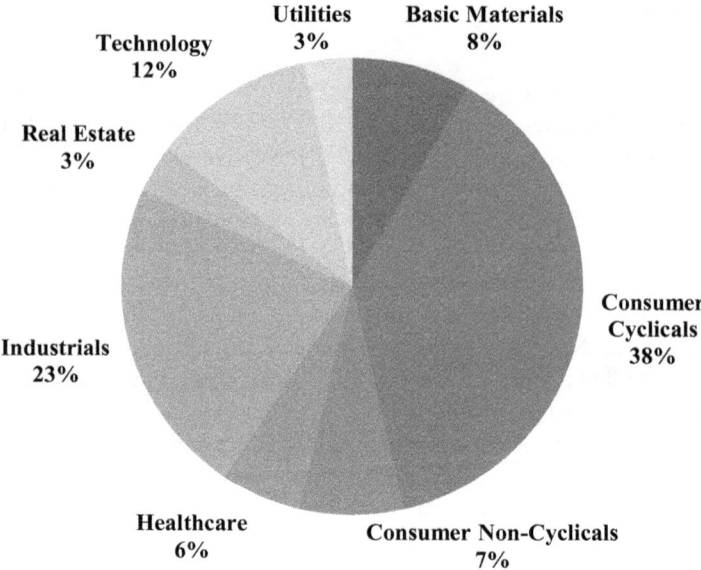

Fig. 5.1 Distribution of sample by industry. Source: Author's elaboration on LSEG workspace data (Accessed July 21, 2024)

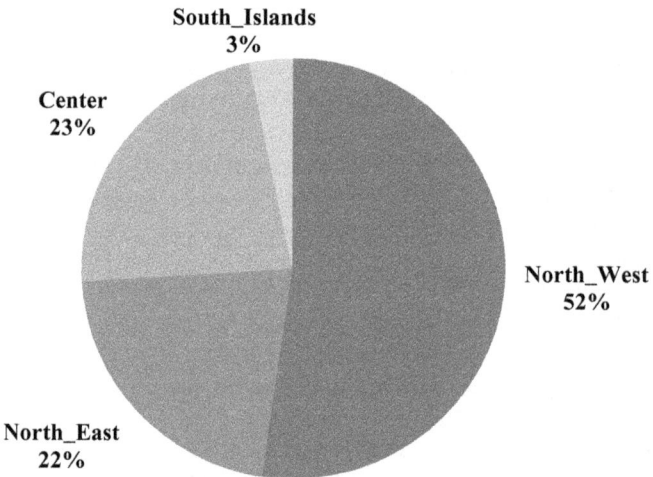

Fig. 5.2 Geographical distribution of sample. Source: Author's elaboration on Registro Imprese's data (Accessed July 21, 2024)

5.4 Comparative Analysis of Governance and Sustainability Practices

5.4.1 Corporate Governance Characteristics

Considering the entire sample of 140 Italian listed family firms, we analyzed the following corporate governance characteristics: board size, board independence, and CEO chairperson duality. We also performed a separate analysis on the subsamples of the 83 companies that are required to disclose nonfinancial information and of the 57 companies that are not subject to this requirement.

5.4.1.1 Board Size

As shown in Fig. 5.3, on average, the boards of directors of Italian listed family firms include nine members. For companies that are subject to nonfinancial disclosure (NFD), the average board consists of ten members, whereas companies that are not subject to NFD have a smaller average board size of seven members. These numbers align with the principles of good governance codes and best practices, which suggest that the optimal board size should range between 9 and 15 directors (Zattoni, 2020). However, considering the two subsamples, companies that are not subject to NFD tend to have fewer board members than those that are subject to NFD. This range is considered optimal, as it allows for effective discussion and decision-making inside the boardroom and ensures contributions from diverse

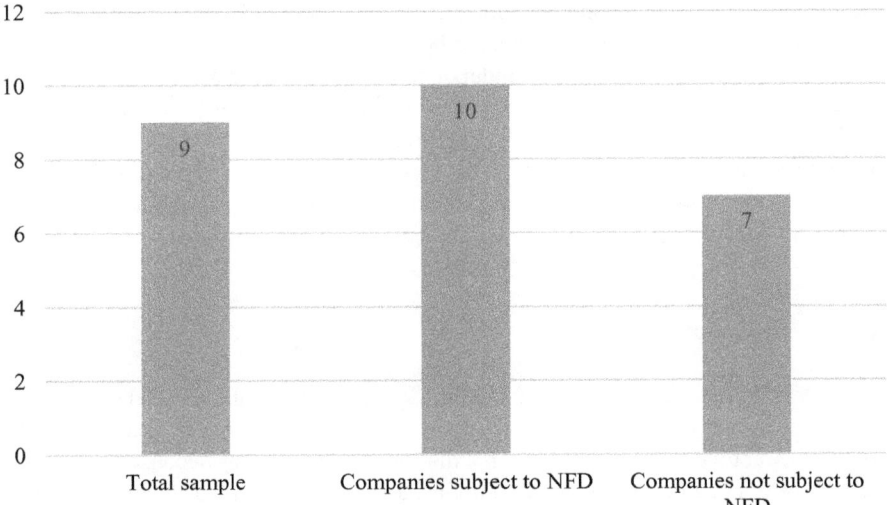

Fig. 5.3 Corporate Governance Characteristics of Family Firms: Board Size. Source: Author's elaboration on data from Borsa Italiana and LSEG Workspace (Accessed July 21, 2024)

perspectives. In fact, a board that is too small may struggle with a lack of varied viewpoints, which may limit the breadth of ideas during boardroom discussions. Conversely, an excessively large board can lead to difficulties in coordination and communication and increased conflicts, potentially resulting in diluted accountability and ineffective decision-making processes. Having a balanced board size enables members to engage in significant debates and collectively leverage their individual expertise (Hillman & Dalziel, 2003; Kiel & Nicholson, 2003). Additionally, different perspectives are essential for addressing complex business challenges and making informed strategic decisions that align with the company's goals. A well-structured board promotes a culture of collaboration and open dialog among directors, which is fundamental for navigating the dynamic and complex business environment. Finally, an optimal board size contributes to improved organizational performance and governance, reinforcing stakeholder trust (Radu et al., 2022).

5.4.1.2 Board Independence

Corporate governance codes emphasize the importance of having a board composed of independent directors, as this composition is crucial for increasing objectivity and transparency in decision-making processes. Independent directors bring diverse perspectives and expertise, which can mitigate potential conflicts of interest and promote better governance practices (Hillman & Dalziel, 2003). Furthermore, a board with a sufficient number of independent members can provide more effective oversight of management, ensuring that the interests of all shareholders are considered. This alignment with best practices not only increases investor confidence but also contributes to the long-term sustainability and performance of the company (De Villiers et al., 2011; Dunn & Sainty, 2009; Hussain et al., 2018).

Figure 5.4 shows that, on average, the boards of directors of Italian domestically listed family firms consist of 45% independent board members. This percentage is consistent across both companies that are subject to NFD (45%) and those that are not subject to such disclosure (44%). The high proportion of independent directors in Italian domestically listed family firms signals a commitment to maintaining rigorous corporate governance standards that align with both international and national norms and best practices.

5.4.1.3 CEO Duality

According to corporate governance codes, the separation of the CEO and the chairperson roles is considered a best practice for promoting independence within the board of directors (Zattoni, 2020). This distinction is essential because it mitigates the risk of concentrated power in a single individual, which can lead to potential conflicts of interest and undermine the board's oversight responsibilities, which is consistent with the principles of agency theory (Fama & Jensen, 1983). By having separate individuals in these roles, the board can benefit from a more objective

5.4 Comparative Analysis of Governance and Sustainability Practices 77

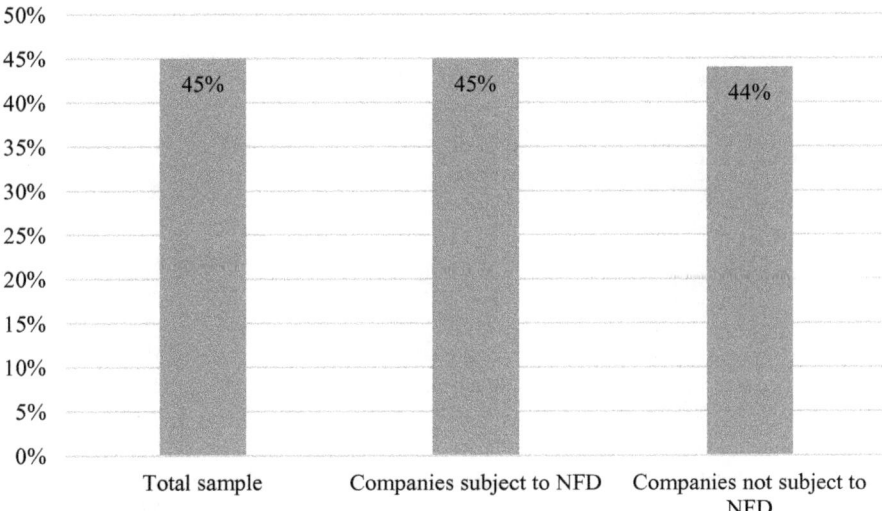

Fig. 5.4 Corporate Governance Characteristics of Family Firms: Independent Board Members. Source: Author's elaboration on data from Borsa Italiana and LSEG Workspace (Accessed July 21, 2024)

perspective, which favors greater accountability and better decision-making processes (Daily & Dalton, 1993). This separation of the roles of running the firm and running the board also encourages a healthy balance of power and diverse viewpoints, contributing to more effective governance and oversight and improved organizational performance, in line with the increasing request of greater transparency by external stakeholders.

As shown in Fig. 5.5, on average, 31% of the boards of directors of Italian domestically listed family firms exhibit CEO chairperson duality. Considering the two subsamples, this percentage is consistent across both companies that are subject to NFD and those that are not subject to it. This may signal reduced independence and oversight ability of boards; therefore, the Italian Corporate Governance Code (Codice di Autodisciplina) recommends appointing a lead independent director when the CEO and chairperson roles are combined to provide a sort of counterbalance and increase the effectiveness of board decision-making.

5.4.2 Sustainability Characteristics

We now examine the sustainability practices and performance of Italian domestically listed family firms. However, owing to missing data, only a subset of 50 companies has available information regarding their sustainability commitments and practices, among which 47 companies are required to disclose nonfinancial information and only 3 companies are not subject to this requirement.

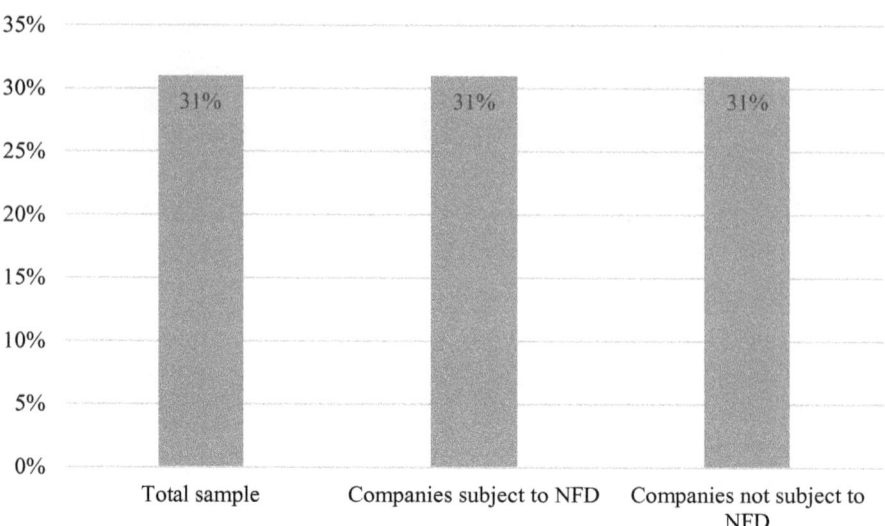

Fig. 5.5 Corporate Governance Characteristics of Family Firms: CEO Duality. Source: Author's elaboration on data from Borsa Italiana and LSEG Workspace (Accessed July 21, 2024)

5.4.2.1 Environmental, Social, and Governance Performance

The ESG score serves as a proxy for a company's commitment to sustainability and is used to measure the company's ESG performance. This score is calculated via LSEG Workspace via publicly accessible data reported by companies, covering ten individual categories that are grouped into three main pillars: environmental, social, and governance. The overall ESG score is derived from these aggregated pillars and is ranked by percentile, benchmarked against the Thomson Reuters industry classification (LSEG Data & Analytics, 2023). The score can range from 0 to 100, with higher scores reflecting greater commitment to ESG practices.

As shown in Fig. 5.6, on average, the companies in the sample have an overall ESG score that is above the mean (i.e., 56), indicating good relative ESG performance and a high degree of transparency in disclosing material ESG data to the public. Specifically, the social pillar presented the highest score among the three (i.e., 63), indicating greater attention to the social dimension than to the other two dimensions.

By analyzing the subsamples of companies that are subject to NFD and those that are not subject to such disclosure, we discover that the former shows higher scores in all categories: the overall ESG score and the three main pillars—environmental, social, and governance.

5.4.2.2 Environmental Performance

An examination of the environmental pillar and its three subpillars—resource use score, emission score, and environmental innovation score (Fig. 5.7)—reveals that the companies in the sample have an environmental score that is just above the mean

5.4 Comparative Analysis of Governance and Sustainability Practices

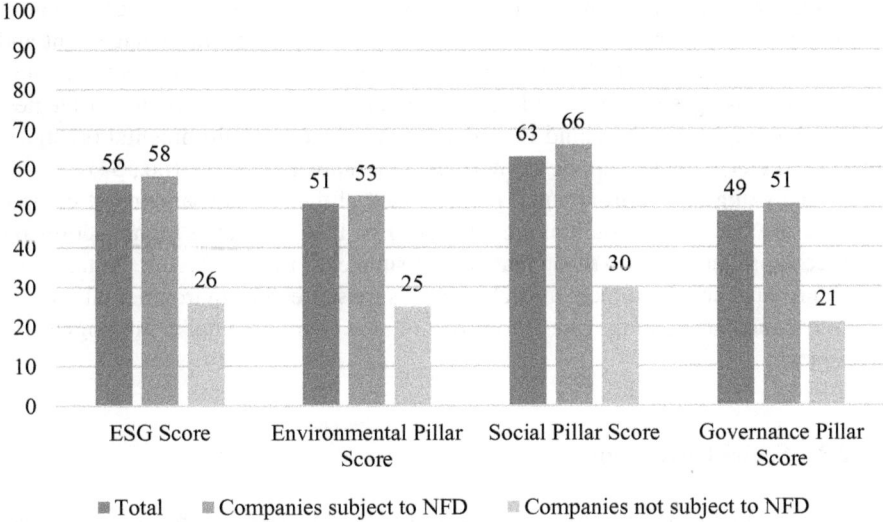

Fig. 5.6 Sustainability Characteristics of Family Firms: ESG Performance. Source: Author's elaboration on LSEG workspace data (Accessed July 21, 2024)

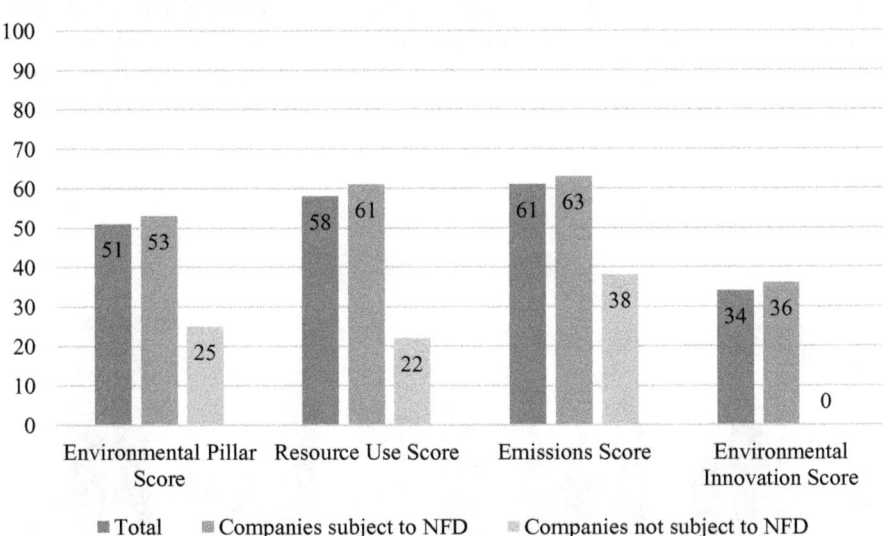

Fig. 5.7 Sustainability Characteristics of Family Firms: Environmental Pillar. Source: Author's elaboration on LSEG workspace data (Accessed July 21, 2024)

(i.e., 51). This indicates a moderate commitment to addressing environmental concerns and reflects a restrained approach to sustainability practices related to the environment. Interestingly, while considering the subpillars, there are higher results for the resource use score (i.e., 58) and the emission score (i.e., 61), suggesting that these companies not only comply with regulatory requirements (e.g., in terms of emissions) but also are committed to increase their environmental performance. Their focus on resource efficiency to improve their supply chain management and on emissions reduction in the production and operational processes seems to suggest an increasing significance of sustainability within their corporate strategies. Although companies lag behind in terms of innovation in environmental practices, they do, in fact, achieve a modest environmental innovation score (i.e., 34).

By analyzing the environmental performance of the two subsamples of companies, we discover that those that are subject to NFD demonstrate higher scores for all three subpillars than do those that are not subject to this disclosure. Notably, for companies that are not subject to such a disclosure score, the environmental innovation score indicates that they need to invest in new environmental technologies and processes.

5.4.2.3 Social Performance

An analysis of the social pillar and its four subpillars—workforce score, human rights score, community score, and product responsibility score (Fig. 5.8)—reveal that companies achieve good social performance. Specifically, companies exhibit a high ranking in the workforce score (i.e., 73), indicating that they are committed to job satisfaction, a healthy and safe workplace, and to developing opportunities for their workforce. Companies also show good results in terms of the product responsibility score (i.e., 62), which signals their commitment to produce high-quality goods and services that protect the customer's health and safety, integrity and data

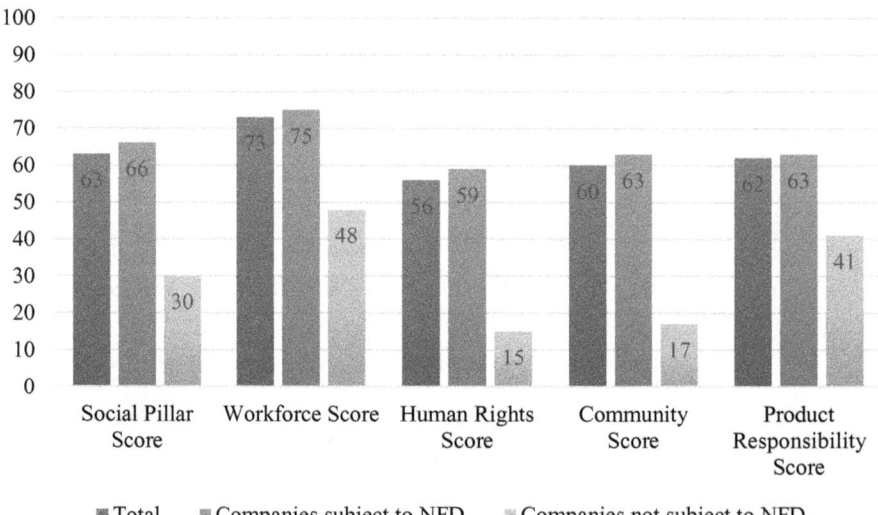

Fig. 5.8 Sustainability Characteristics of Family Firms: Social Pillar. *Source*: Author's elaboration on LSEG workspace data (Accessed July 21, 2024)

privacy, and in the community score (i.e., 60), which highlights the firm's commitment to being a responsible corporate citizen and assesses how well a company engages with local communities, contributing to protecting public health and respecting ethical standards in its business conduct. Companies have good human rights scores (i.e., 56), indicating their adherence to fundamental human rights conventions and standards. This metric generally reflects the presence of a company's policies and practices that promote and safeguard human rights within companies' operations and supply chains.

Considering the social performance of the two subsamples of companies, those that are subject to NFD consistently achieve higher social scores for all four subpillars than do those that are not subject to this disclosure. Notably, the most pronounced disparities are found in the human rights score and in the community score.

5.4.2.4 Governance Performance

As shown in Fig. 5.9, an examination of the governance pillar and its three subpillars—management score, shareholder score, and CSR strategy score—shows that companies have achieved a score that is approximately average (i.e., 49), indicating adequate governance performance.

This can be first observed in the management score (i.e., 49), which signals that companies have a sufficient commitment to and effectiveness in adhering to best corporate governance practices. Additionally, the shareholders score (i.e., 51) indicates good discreet use of practices regarding shareholder equality and the use of

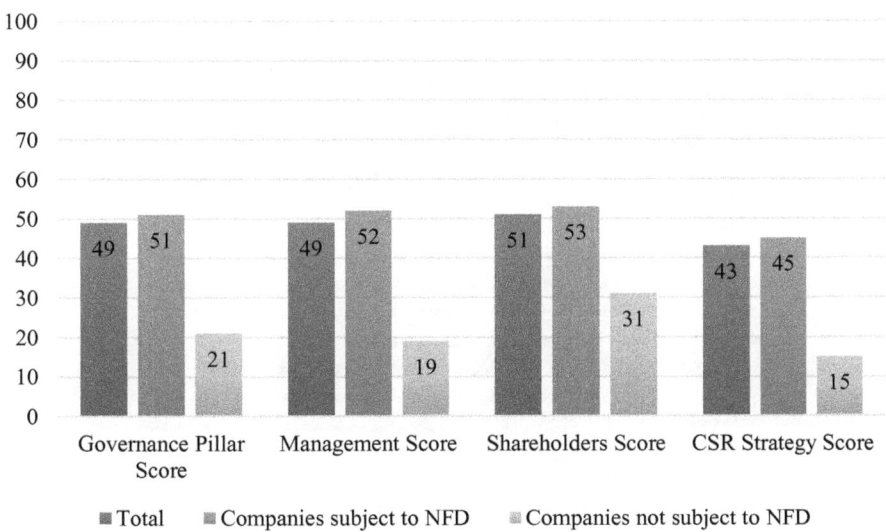

Fig. 5.9 Sustainability Characteristics of Family Firms: Governance Pillar. *Source*: Author's elaboration on LSEG workspace data (Accessed July 21, 2024)

mechanisms to prevent takeovers by companies. However, companies show lower results in the CSR strategy score (i.e., 43), which reflects how well a company communicates its integration of economic, social, and environmental considerations into its daily decision-making processes. This highlights the need for companies to prioritize sustainability aspects into their corporate strategies to position themselves as leaders in responsible business practices.

An analysis of the governance performance of the two subsamples of companies reveals that those that are subject to NFD have higher scores for all three subpillars than do those that are not subject to this disclosure. The most significant differences can be observed in the management score, followed by the CSR strategy score, highlighting the importance of companies having a greater commitment to sustainability.

5.4.2.5 Sustainable Development Goals (SDGs)

We also need to examine the role of family firms in contributing to the SDGs (United Nations Sustainable Development Goals, 2015). As shown in Fig. 5.10, the most actively pursued goals are decent work and economic growth (SDG 8), good health and well-being (SDG 3), and responsible consumption and production (SDG 12), followed by climate action (SDG 13), affordable and clean energy (SDG 7), and gender equality (SDG 5). By actively pursuing these goals, family firms can leverage their unique attributes—such as long-term vision, community orientation, and stakeholder engagement—to create sustainable value.

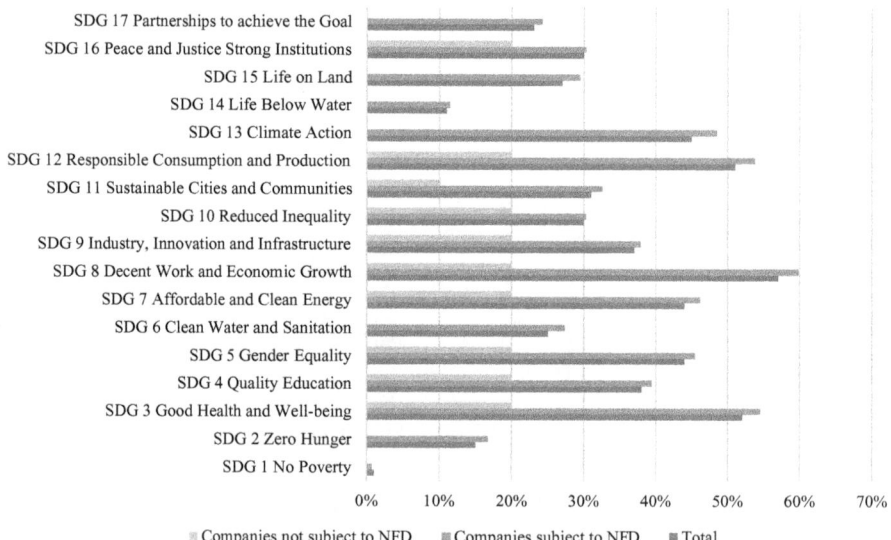

Fig. 5.10 Sustainability Characteristics of Family Firms: SDGs. *Source*: Author's elaboration on LSEG workspace data (Accessed July 21, 2024)

5.4 Comparative Analysis of Governance and Sustainability Practices

First, companies are engaged in promoting sustainable economic growth, creating employment opportunities, and ensuring decent working conditions for all, as indicated by SDG 8. Companies also seem to be involved in the social dimension. For example, to contribute to good health and well-being (SDG 3), they can prioritize employee health by implementing robust safety measures or by engaging in community health initiatives to demonstrate corporate responsibility. Similarly, by prioritizing SDG 5, companies can adopt more inclusive workplace policies that attract diverse talent, thus promoting a culture of equality and respect. Moreover, companies address environmental concerns. Specifically, by focusing on SDG 12, they can develop more sustainable production processes that minimize waste and promote resource efficiency; they can also take actions to fight climate change in line with SDG 13 climate action and to guarantee access to affordable and clean energy, as stated by SDG 7.

In contrast, the least pursued goals are no poverty (SDG 1), living below water (SDG 14), and zero hunger (SDG 2). This indicates the need for companies to implement more initiatives in pursuit of these significant objectives, therefore contributing to the end of hunger and poverty in all its forms and preventing marine pollution.

Examining the two subsamples of companies that are subject to NFD and those that are not subject to such disclosure, it is evident that the former demonstrates a high degree of engagement with the SDGs.

5.4.2.6 Corporate Policies Related to Sustainability

Figure 5.11 illustrates the presence of corporate policies related to sustainability in Italian domestically listed family firms. Companies have established policies across all three domains—environmental, social, and governance. Specifically, they demonstrate a strong focus on the social dimension, specifically with respect to employee health and safety and diversity and opportunity. Additionally, these companies have codes of conduct that outline their commitment to upholding the highest standards of business ethics and integrity and to avoid bribery and corruption at all their operations. Companies also demonstrate a high level of stakeholder engagement. Notably, they provide information on how companies engage with their stakeholders, how they involve them in their decision-making processes, and what procedures are in place for their engagement. Moreover, companies are highly dedicated to the environmental dimension: they establish processes to efficiently use energy in their operation, and they are involved in lessening the environmental impact of their supply chain and in controlling and reducing emissions in their operations. However, less than half of the sample indicates that their compensation policies include remuneration for the CEO, executive directors, nonboard executives, and other management bodies on the basis of ESG or sustainability factors.

Considering the two subsamples of companies, those that are subject to NFD show high results both in corporate policies related to the social dimension and to the environmental dimension.

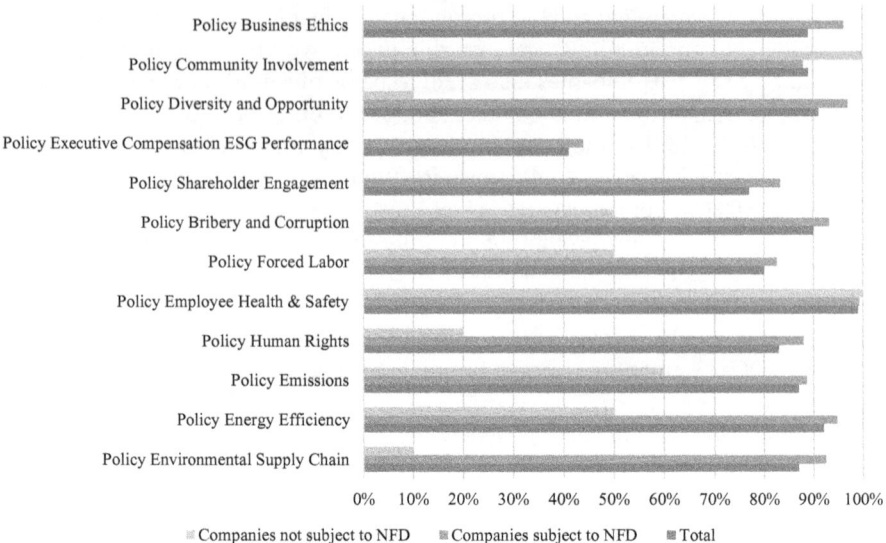

Fig. 5.11 Sustainability Characteristics of Family Firms: Corporate Policies Related to Sustainability. Source: Author's elaboration on LSEG workspace data (Accessed July 21, 2024)

5.4.2.7 Sustainability and Governance in Family Firms

By examining the corporate governance characteristics of the sample with available sustainability information, we focus on the board of directors and its committees.

Specifically, companies that are subject to NFD have, on average, a board composed of 11 members (which is almost double that of companies that are not subject to NFD), with an average tenure of board members of 9 years (Fig. 5.12). These findings highlight that a larger and more experienced board may be better equipped to integrate sustainability considerations into corporate strategy and decision-making processes.

Regarding the characteristics of boards of directors, as shown in Fig. 5.13, an examination of board composition reveals that companies that are subject to NFD have a high presence of independent directors and a moderate number of executive directors, and they demonstrate greater gender diversity. A large representation of independent directors not only increases board effectiveness by introducing diverse perspectives and expertise but also strengthens the ability of boards to prioritize sustainability initiatives. These directors can challenge other directors' views and advocate for long-term strategies that align with ESG goals. Moreover, greater gender diversity contributes to a more inclusive decision-making process in the boardroom, which is also critical for promoting innovative approaches to sustainability challenges. Diverse boards are better positioned to understand the different needs of their stakeholders, particularly those related to sustainable practices. By integrating these diverse viewpoints, companies can better navigate the complexities of

5.4 Comparative Analysis of Governance and Sustainability Practices

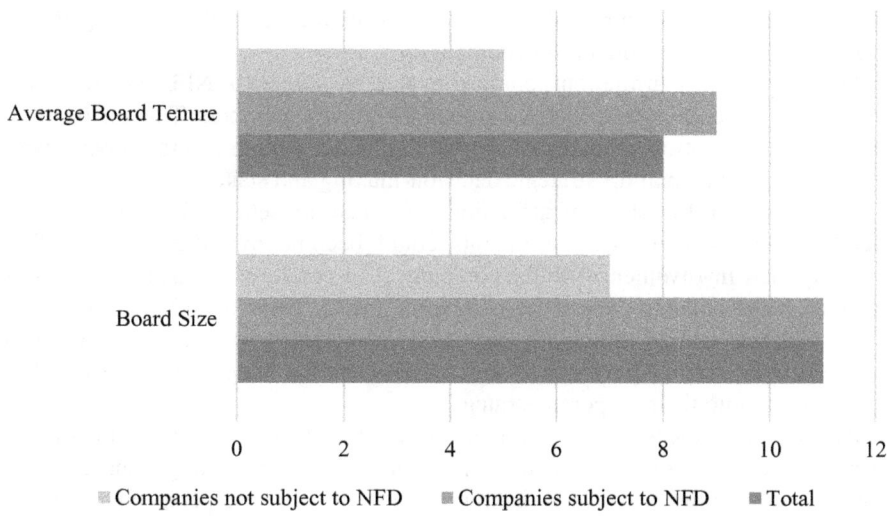

Fig. 5.12 Sustainability and Governance in Family Firms: Board of Directors. Source: Author's elaboration on data from Borsa Italiana and LSEG Workspace (Accessed July 21, 2024)

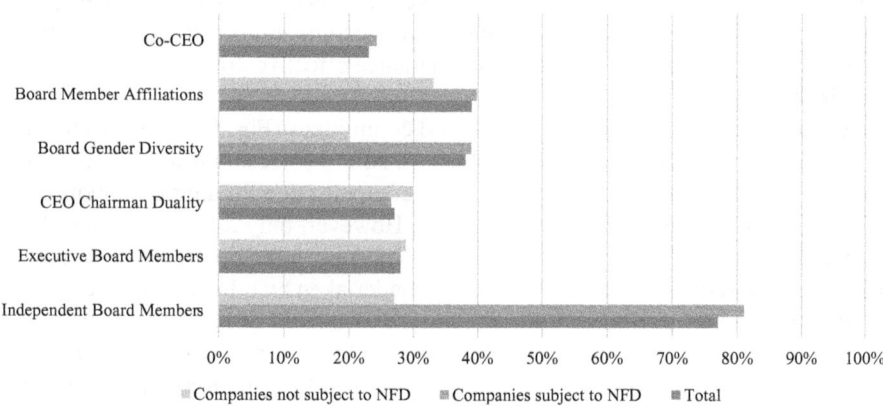

Fig. 5.13 Sustainability and Governance in Family Firms: Board of Directors' Characteristics. Source: Author's elaboration on data from Borsa Italiana and LSEG Workspace (Accessed July 21, 2024)

sustainability and implement effective strategies that contribute to both corporate performance and positive societal impact.

Additionally, on average, 31% of the boards of directors of companies that are subject to NFD exhibit CEO chairperson duality. While this arrangement can enable streamlined decision-making, it concentrates power within a single individual, potentially undermining the board's oversight capabilities. This concentration of power could hinder the board's effectiveness in addressing sustainability challenges and integrating them into corporate strategy. Therefore, family firms must assess the implications of maintaining this governance structure carefully, ensuring that it aligns with their commitment to sustainability.

In terms of board affiliations, companies that are subject to NFD tend to have a high number of corporate affiliations among their board members. This can be interpreted as an indication of increased networking and access to diverse perspectives, which can be beneficial for strategic decision-making and stakeholder engagement. However, this proliferation of affiliations may also present challenges, as board members with numerous commitments could become overwhelmed, therefore reducing their involvement with the company. As a result, while these connections can enrich the collective expertise of boards, family firms must verify that their board members remain sufficiently dedicated to their roles, as a lack of engagement may negatively impact governance effectiveness and the integration of sustainability initiatives into their corporate strategy.

An analysis of the leadership structure reveals that nearly a quarter of the sample operates with co-CEOs at the helm of the organization. This arrangement can contribute to collaborative decision-making and leverage diverse perspectives, which may increase family firms' ability to integrate sustainability initiatives into their strategies. By combining their strengths, co-CEOs can implement sustainability efforts more effectively, ensuring that ESG considerations are embedded in the company's strategies and operations. However, this shared leadership structure may also bring challenges related to authority and accountability, so roles need to be clearly communicated and delineated to guarantee effective governance in pursuing sustainability goals.

An examination of the presence of board committees (Fig. 5.14) reveals that the vast majority of companies that are subject to NFD have established the following board committees: nomination committees, compensation committees, audit committees, and CSR sustainability committees. However, only a few companies have dedicated corporate governance committees. This may indicate either that its role of advising on corporate governance matters is undertaken by other committees or that companies need to exert more efforts to provide oversight of good governance.

Specifically, audit committees, which are responsible for overseeing financial reporting and internal controls and for compliance with regulations, including those related to sustainability, play an important role. Therefore, audit committees hold a critical position in the oversight of sustainability disclosure, in line with the growing demand for this information by policymakers and regulators.

Nomination committees, which are responsible for evaluating a firm's board of directors and identifying candidates for positions on a board on the basis of the skills and characteristics needed, play a key role in providing companies with sustainability capabilities. Hence, the role of these committees is particularly important when considering the recruitment of members with sustainability experience from outside family firms.

Compensation committees are in charge of determining executive compensation packages and should be composed of independent directors, according to the Corporate Governance Code. Companies are increasingly incorporating ESG metrics, assessed on either a quantitative or a qualitative basis, into executive

5.4 Comparative Analysis of Governance and Sustainability Practices

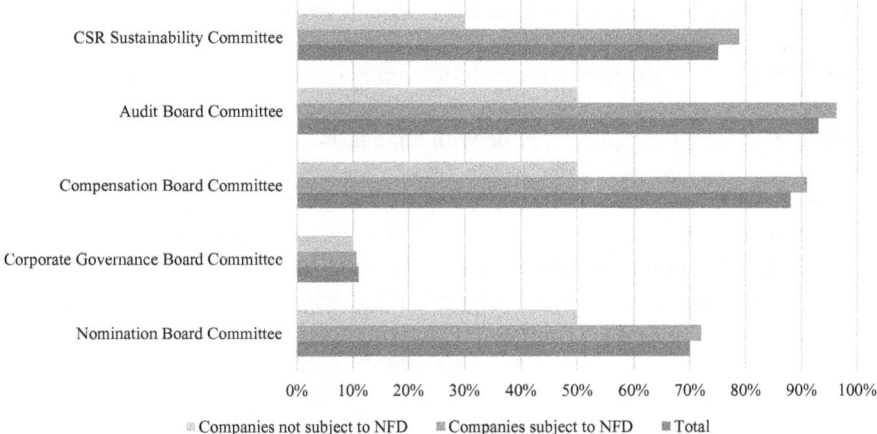

Fig. 5.14 Sustainability and Governance in Family Firms: Board Committees. Source: Author's elaboration on data from Borsa Italiana and LSEG Workspace (Accessed July 21, 2024)

compensation plans. Thus, compensation committees may contribute significantly to promoting sustainability in family firms, guaranteeing transparency in executive compensation to build stakeholder trust.

The presence of a CSR sustainability committee in family firms is a positive indicator of a company's commitment to CSR and sustainable practices. This dedicated committee not only oversees the implementation of sustainability initiatives but also ensures that ESG factors are integrated into the company's strategic decision-making processes. By establishing such a committee, family firms demonstrate their willingness to engage with stakeholders, assess their impact on society and the environment, and continuously improve their sustainability efforts. Moreover, the presence of this committee can signal transparency and commitment to sustainability to external stakeholders, reinforcing a company's reputation and building stakeholder trust.

5.5 Best Practices and Lessons Learned

The above analysis of a sample of Italian listed family firms has provided valuable insights into the landscape of corporate governance characteristics and sustainability practices that can effectively support the implementation of environmental, social, and governance initiatives, ultimately leading to positive sustainability outcomes.

From a governance perspective, the establishment of a CSR sustainability committee can significantly enhance a company's ability to integrate sustainability matters into the board's agenda. Such a committee serves as a dedicated body that ensures that sustainability issues are included during board discussions, encouraging accountability regarding sustainability initiatives among board members.

By providing a structured approach to evaluating and addressing a company's environmental and social impacts, a CSR sustainability committee can contribute to the development of effective sustainability strategies and policies. Additionally, it can help position sustainability as a fundamental aspect of a company's strategy rather than merely a compliance issue with regulations. This proactive approach can improve relationships with stakeholders and increase the family firm's reputation.

In terms of board composition, having sufficient board members can broaden the range of perspectives and bring diverse expertise in the boardroom, enriching discussions, and decision-making processes. Additionally, having several independent directors can significantly enhance a board's objectivity and effectiveness, ensuring that it remains accountable to all stakeholders. By representing the interests of various stakeholder groups—including shareholders, employees, customers, and the community—independent directors play a crucial role in promoting sustainable practices within the company, strengthening the board's ability to address new complex challenges.

Several best practices for advancing sustainability are worth highlighting. First, establishing clear corporate policies related to sustainability represents an effective measure to commit companies to sustainability. These policies should outline specific goals, responsibilities, and performance metrics that align with ESG principles. Additionally, engaging stakeholders in the development and implementation of these policies can increase their effectiveness. Regular training for employees on sustainability practices and integrating sustainability into a company's strategy can further reinforce this commitment. Moreover, publicly reporting on the sustainability initiatives implemented by the company not only increases trust among stakeholders but also contributes to enhancing the family firm's reputation.

Companies could improve their communication by clearly outlining which SDGs they aim to target, thereby demonstrating their commitment to aligning with global efforts. By explicitly identifying and prioritizing specific SDGs, family firms can clarify their sustainability objectives to stakeholders, enabling a deeper understanding of their contributions to global challenges. Furthermore, transparent reporting on the progress made toward these goals can increase credibility with investors and trust with customers, suppliers, and the community, also encouraging collaboration with other organizations working toward similar objectives. In fact, engaging in partnerships and sharing best practices related to these SDGs can increase family firms' impact and even inspire others in the industry to take meaningful action in an ever-evolving landscape.

However, the implementation of environmental, social, and governance practices presents a multifaceted set of challenges for family firms. Financial costs remain a primary obstacle, as investments in sustainable technologies, processes, or supply chains can be highly expensive, especially when short-term financial returns are uncertain. Beyond the financial dimension, the increasing complexity of regulatory frameworks places significant administrative and compliance burdens, often requiring higher levels of professionalization, more formalized internal governance, and structured data collection and reporting mechanisms, which can affect the established routines and informal processes typical of many family firms. Resistance to change is another critical barrier, as sustainable initiatives may challenge consolidated practices and value systems, particularly in family firms where tradition, continuity, and intergenerational control are central. This resistance may be exacerbated by generational tensions, where younger family members advocate for sustainability integration while older leaders remain cautious or skeptical about its strategic or financial relevance. Another significant constraint is the frequent lack of internal ESG expertise, which undermines family firms' ability to design and implement effective sustainability strategies. Additionally, many family businesses operate in sectors or markets where pressure from external stakeholders, such as customers or investors, may be less pronounced, reducing the perceived urgency for transformation.

Taken together, these challenges create a complex landscape in which advancing sustainability within family firms requires not only adequate financial and human resources but also a cultural shift, a renewed leadership mindset, and a long-term strategic vision. Although seeking external expertise through consultants or partnerships can support the development of effective sustainability strategies, significant progress may also depend on overcoming internal skepticism, favoring dialogue across generations, and aligning ESG objectives with the family's identity and values.

References

Amore, M. D., Minichilli, A., & Corbetta, G. (2011). How do managerial successions shape corporate financial policies in family firms? *Journal of Corporate Finance, 17*(4), 1016–1027. https://doi.org/10.1016/j.jcorpfin.2011.05.002

Anderson, R. C., & Reeb, D. M. (2003). Founding-family ownership and firm performance: Evidence from the S&P 500. *The Journal of Finance, 58*(3), 1301–1328. https://doi.org/10.1111/1540-6261.00567

Barth, E., Gulbrandsen, T., & Schønea, P. (2005). Family ownership and productivity: The role of owner-management. *Journal of Corporate Finance, 11*(1–2), 107–127. https://doi.org/10.1016/j.jcorpfin.2004.02.001

Borsa Italiana. (2024). *Corporate governance reports*. Retrieved July 20, 2024, from https://www.borsaitaliana.it/borsa/azioni/documenti/societa-quotate/governance-societa-quotate.html

CONSOB. (2024). Retrieved July 20, 2024, from https://www.consob.it/web/consob-and-its-activities

Daily, C. M., & Dalton, D. R. (1993). Board of directors leadership and structure: Control and performance implications. *Entrepreneurship Theory and Practice, 17*(3), 65–81. https://doi.org/10.1177/104225879301700305

De Villiers, C., Naiker, V., & van Staden, C. J. (2011). The effect of board characteristics on firm environmental performance. *Journal of Management, 37*(6), 1636–1663. https://doi.org/10.1177/0149206311411506

Dunn, P., & Sainty, B. (2009). The relationship among board of director characteristics, corporate social performance and corporate financial performance. *International Journal of Managerial Finance, 5*(4), 407–423. https://doi.org/10.1108/17439130910987558

EU Non-Financial Reporting Directive. (2014). Retrieved July 3, 2024, from https://eur-lex.europa.eu/legal-content/EN/TXT/?uri=celex%3A32014L0095

Fama, E. F., & Jensen, M. C. (1983). Separation of ownership and control. *The Journal of Law and Economics, 26*(2), 301–325. https://doi.org/10.1086/467037

Gazzetta Ufficiale della Repubblica Italiana. (2016). *Italian Legislative Decree No. 254/2016*. Retrieved July 3, 2024, from https://www.gazzettaufficiale.it/eli/id/2017/01/10/17G00002/sg

Hillman, A. J., & Dalziel, T. (2003). Boards of directors and firm performance: Integrating agency and resource dependence perspectives. *Academy of Management Review, 28*(3), 383–396. https://doi.org/10.5465/amr.2003.10196729

Hussain, N., Rigoni, U., & Orij, R. P. (2018). Corporate governance and sustainability performance: Analysis of triple bottom line performance. *Journal of Business Ethics, 149*(2), 411–432. https://doi.org/10.1007/s10551-016-3099-5

Kiel, G. C., & Nicholson, G. J. (2003). Board composition and corporate performance: How the Australian experience informs contrasting theories of corporate governance. *Corporate Governance: An International Review, 11*(3), 189–205. https://doi.org/10.1111/1467-8683.00318

LSEG (London Stock Exchange Group) Workspace. (2024). Retrieved July 21, 2024, from https://workspace.refinitiv.com/web

LSEG Data & Analytics. (2023). *Environmental, social and governance scores from LSEG*. Retrieved July 21, 2024, from https://www.lseg.com/content/dam/data-analytics/en_us/documents/methodology/lseg-esg-scores-methodology.pdf

Radu, C., Smaili, N., & Constantinescu, A. (2022). The impact of the board of directors on corporate social performance: A multivariate approach. *Journal of Applied Accounting Research, 23*(5), 1135–1156. https://doi.org/10.1108/jaar-05-2021-0141

Registro Imprese. (2024). *Official data from the Chambers of Commerce*. Retrieved July 1, 2024, from https://www.registroimprese.it/

United Nations Sustainable Development Goals. (2015). Retrieved July 1, 2024, from https://sdgs.un.org/goals

Zattoni, A. (2020). *Corporate governance. How to design good companies*. Bocconi University Press.

Chapter 6
Conclusion

6.1 Summary of Key Findings

By empirically examining the governance mechanisms and sustainability practices of family firms in the Italian context, this book contributes to the growing body of literature on family business management, offering a nuanced understanding of how the distinctive characteristics of family firms shape their approach to sustainability, thereby advancing theoretical and practical insights in this evolving field.

Given the pressing global sustainability challenges such as climate change and social responsibility, it is now more crucial than ever to understand how to encourage sustainable practices. The pursuit of sustainability is not merely a challenge; it represents a significant opportunity for family businesses to demonstrate their commitment to long-term success.

As highlighted by the empirical analysis, since the EU enacted the NFRD, which compels companies to disclose nonfinancial information on ESG matters (EU Non-Financial Reporting Directive, 2014), Italian family firms have demonstrated an increasing level of commitment to sustainability. This commitment is evident in their positive ESG performance, which not only aligns with regulatory requirements but also reflects a deeper understanding of the importance of sustainability in creating long-term value.

Global initiatives such as the SDGs of the UN (United Nations Sustainable Development Goals, 2015), together with the regulatory impulse, have served as catalysts for these firms, encouraging them to integrate sustainability considerations into their strategies and governance structures. By actively reporting on their ESG initiatives, these companies increase transparency and build trust among their stakeholders. Furthermore, an active approach to sustainability and greater stakeholder engagement adopted by Italian family firms can position them favorably in a competitive market that is increasingly driven by environmental, social, and governance considerations, allowing them to differentiate themselves as responsible corporate citizens. This positive trend indicates that family firms' adherence to regulatory frameworks may represent not merely a compliance exercise but also a strategic advantage that can lead family firms to improved business outcomes.

6.2 Implications for Family Businesses

Owing to their unique features and values (Astrachan et al., 2002; Berrone et al., 2012; Eddleston & Kellermanns, 2007; Tagiuri & Davis, 1996), family firms can cultivate a shared sense of purpose and responsibility, driving collective organizational efforts toward achieving sustainable outcomes that resonate with both internal family firms' values and external stakeholders' expectations.

Instituting dedicated governance structures and establishing corporate policies related to sustainability matters position family businesses to thrive in a market in which sustainability is becoming a key driver of competitive advantage.

The implementation of good governance practices serves as a crucial mechanism for promoting sustainability in Italian family businesses. By embedding sustainability into their governance framework, family firms can promote a culture of ethical stewardship inside the company, specifically through engaging employees at all levels to integrate ESG considerations into their daily activities, thus aligning the entire workforce with a company's sustainability goals.

CSR sustainability committees represent an important governance body that advances sustainability and signals a strong commitment to address sustainability issues in boardroom discussions and corporate strategies, guaranteeing that the decisions and initiatives that are undertaken align with a company's overall goals and values.

Additionally, establishing explicit corporate policies related to sustainability can effectively communicate a family firm's commitment to sustainable practices to its diverse stakeholders. As stakeholders increasingly seek assurance that companies are not only compliant with regulations but also genuinely dedicated to making a positive impact, clear policies serve as important signals for transparency. These policies can outline the specific sustainability goals and initiatives pursued by family businesses, therefore, demonstrating their progress in achieving these objectives.

Furthermore, family businesses can greatly benefit from actively engaging stakeholders in the development and implementation of sustainability initiatives. This collaborative approach contributes to developing stronger relationships built on trust and ensures that the initiatives carried out by family firms are relevant and aligned with the needs and expectations of all parties involved. By involving internal and external stakeholders—such as employees, customers, suppliers, and the local community—in sustainability discussions, family firms can collect valuable insights and feedback that increase the effectiveness of their initiatives. Additionally, this engagement can lead to increased support from stakeholders, further amplifying the positive impact of family firms' sustainability efforts.

Finally, family businesses may benefit from linking executive pay to sustainability-related goals to create a strong incentive to prioritize sustainable practices within the organization. By introducing performance metrics tied to ESG outcomes, family businesses ensure that executives are held accountable for achieving specific sustainability objectives. This strategy helps drive both short-term performance and long-term value creation, positively impacting stakeholders' perceptions. In fact, the alignment of financial rewards with sustainability signals to stakeholders that the organization is genuinely committed to long-term sustainability. Moreover, when executives have a vested interest in meeting these goals, this can lead to an increased commitment to sustainable practices and greater investment in initiatives that increase a family firm's overall environmental, social, and governance performance.

6.3 Policy Recommendations

Policymakers can play a pivotal role in helping family businesses be aware of and navigate the complexities of sustainability challenges, ultimately contributing positively to society and the environment. By enacting clear and consistent regulations regarding sustainability reporting and disclosure, policymakers can provide family firms with the guidance that is necessary to align their strategies and operations with broader ESG goals. In fact, it is important not only to assist family businesses in addressing environmental and social issues but also to nurture a culture of sustainability that may provide benefits to the entire community, leading to a more sustainable economy.

Policymakers and regulators can also implement initiatives aimed at promoting sustainable corporate governance in family firms by defining clear requirements and offering actionable recommendations for incorporating sustainability issues into companies' decision-making processes. By establishing comprehensive guidelines that indicate the best practices for integrating ESG factors into corporate governance frameworks, policymakers can contribute to the inclusion of sustainability as a critical aspect of the boardroom agenda of family firms.

Moreover, by developing supportive frameworks—such as tax incentives, grants, and resources specifically dedicated to family firms—policymakers can encourage these businesses to invest in sustainable practices. For example, by providing financial incentives, policymakers can alleviate the initial costs associated with implementing sustainability initiatives, making it more feasible for family businesses to adopt environmentally friendly technologies and practices. These supportive frameworks can also include grants aimed at funding research and development in sustainable practices, allowing family firms to innovate and increase their efficiency.

Additionally, the creation of educational programs, events, and workshops focused on sharing best practices, experiences, and ideas can provide family firms, particularly SMEs, with the knowledge and tools they need to effectively implement sustainability initiatives. Collaborative platforms that facilitate partnerships between family businesses and other organizations, such as government entities and non-profit organizations, can further increase the sharing of best practices and innovative solutions.

Additionally, family firms can benefit from accessing sustainability networks, which may provide essential guidance and support, as they work to integrate sustainability considerations into their operations and supply chain. The creation of these networks facilitates knowledge sharing, allowing family businesses to learn from one another's experiences. Moreover, by creating a community centered on shared sustainability challenges, these networks can stimulate collective problem solving and idea exchange. As a result, policymakers may significantly contribute not only to improving the capabilities of individual family firms but also to strengthening the entire family business sector's commitment to sustainability initiatives, ultimately generating positive impacts for all stakeholders.

6.4 Final Thoughts

New global scenarios are reshaping the landscape of business, so traditional corporate strategy and governance must be reevaluated. In this evolving context, family firms are increasingly called upon to address sustainability challenges with increased commitment.

These firms have a unique advantage, as their deep-rooted family values can serve as a guiding framework for integrating sustainability into their strategies and operations. By prioritizing sustainability, family firms can leverage their inherent strengths—such as strong interpersonal relationships, a long-term vision, and a commitment to community well-being—to promote an inclusive approach to long-term value creation for all diverse stakeholders. This focus not only benefits the environment and society but also enhances their competitiveness in a market that increasingly requires responsible business practices. Emphasizing sustainability enables family firms not only to mitigate risks but also to actively support transformative actions, all of which are grounded in a strong set of values and cultural principles, ultimately leading to long-term success.

As these firms navigate the complexities of grand sustainability challenges, their commitment to family values and sustainable practices can position them as leaders in creating lasting, positive change.

References

Astrachan, J. H., Klein, S. B., & Smyrnios, K. X. (2002). The F-PEC scale of family influence: A proposal for solving the family business definition problem. *Family Business Review, 15*(1), 45–58. https://doi.org/10.1111/j.1741-6248.2002.00045.x

References

Berrone, P., Cruz, C., & Gomez-Mejia, L. R. (2012). Socioemotional wealth in family firms: Theoretical dimensions, assessments approaches and agenda for future research. *Family Business Review, 25*(3), 258–279. https://doi.org/10.1177/0894486511435355

Eddleston, K. A., & Kellermanns, F. W. (2007). Destructive and productive family relationships: A stewardship theory perspective. *Journal of Business Venturing, 22*(4), 545–565. https://doi.org/10.1016/j.jbusvent.2006.06.004

EU Non-Financial Reporting Directive. (2014). Retrieved July 3, 2024, from https://eur-lex.europa.eu/legal-content/EN/TXT/?uri=celex%3A32014L0095

Tagiuri, R., & Davis, J. (1996). Bivalent attributes of the family firm. *Family Business Review, 9*(2), 199–208. https://doi.org/10.1111/j.1741-6248.1996.00199.x

United Nations Sustainable Development Goals. (2015). Retrieved July 1, 2024, from https://sdgs.un.org/goals

GPSR Compliance

The European Union's (EU) General Product Safety Regulation (GPSR) is a set of rules that requires consumer products to be safe and our obligations to ensure this.

If you have any concerns about our products, you can contact us on

ProductSafety@springernature.com

In case Publisher is established outside the EU, the EU authorized representative is:

Springer Nature Customer Service Center GmbH
Europaplatz 3
69115 Heidelberg, Germany

www.ingramcontent.com/pod-product-compliance
Lightning Source LLC
Chambersburg PA
CBHW070251041125
34948CB00003B/65